THE SOUTHWEST HISTORICAL SERIES

EDITED BY

LeROY R. HAFEN

X

MANHATTAN IN 1860
Showing emigrant train

COLORADO GOLD RUSH

Contemporary Letters
and Reports
1858-1859

edited by

LeROY R. HAFEN, PH.D., LITT.D.
Historian of the State Historical Society of Colorado
Author of The Overland Mail, Colorado, Fort Laramie, etc.

PORCUPINE PRESS
Philadelphia
1974

First edition 1941
(Glendale: The Arthur H. Clark Co., 1941)

Reprinted 1974 by
PORCUPINE PRESS, INC.
Philadelphia, Pennsylvania 19107

Library of Congress Cataloging in Publication Data

Hafen, Le Roy Reuben, 1893–   ed.
    Colorado gold rush: contemporary letters and
reports, 1858–1859.

    (The Southwest historical series, 10)
    Reprint of the 1941 ed. published by A. H. Clark
Co., Glendale, Calif.
    1. Colorado—History—To 1876. 2. Gold mines and
mining—Colorado. I. Title.
F786.S752 vol. 10   [F780]   917.9'03'2 [917.88'8'032]
ISBN 0–87991–306–1   74–7163

F
786
.S 752
V.10

Manufactured in the United States of America

62410

To
JAMES F. WILLARD
and
MARGARET WHEELER WILLARD
Whose scholarly research produced basic material
presented in this volume

CONTENTS

ILLUSTRATIONS

PREFACE

PREFACE

Most of the material presented in this volume was gathered by Doctor and Mrs. James F. Willard and by Elmer R. Burkey. The contribution of Mr. Burkey was described in the preface to the preceding volume in this *Southwest Historical Series.*

In addition to his scholarly researches in the field of medieval English history, Professor Willard, of the University of Colorado, found time to gather source material on various phases of Colorado history. During vacation periods for a number of years he and Mrs. Willard searched the newspapers of the country for Pike's Peak gold rush data. Before the gathering was completed, Doctor Willard's work was terminated by death (November 1935). The material gathered was placed by Mrs. Willard in the University of Colorado Historical Collections, a sizable assemblage of historical sources gathered primarily by Professor Willard.

The present editor's study of Pike's Peak gold rush guidebooks led him of course to the contemporary newspapers. The year spent by Mr. Burkey for the State Historical Society of Colorado in searching Missouri valley newspapers for historical data on Colorado, was primarily devoted to other newspapers than those searched by the Willards. Thus the gatherings supplement each other.

Mrs. Margaret Willard kindly expressed a willingness for the present editor to utilize the gold rush data she and her husband had assembled. Professors C. C. Eckhardt and C. B. Goodykoontz of the history department, University of Colorado, and President Robert L. Stearns of the same institution, generously approved the proposal. It was felt that inasmuch as a study of the gold rush to Colorado was being undertaken, it would be well to pool the available resources. A more satisfactory and representative selection of material could then be presented. My sincere thanks go to the above-named persons for their consideration and generosity.

With more letters and reports than could be reproduced in a single volume, it was necessary for the editor to choose only the more important and the typical reports. He was also forced to delete from some of the accounts the duplicate and less pertinent material. The resulting selections should give a satisfactory picture of the gold rush that brought about the founding of Colorado.

For the sake of uniformity, a few punctuation changes have been made and some minor spelling and typographical errors corrected.

INTRODUCTION

INTRODUCTION

The story of the prospecting parties that came to the eastern base of the Rocky mountains in the summer of 1858 has been told in the preceding volume of this *Southwest Historical Series*. Now we present a selected group of contemporary letters and reports reflecting and describing the responses and the developments that resulted from those initial discoveries.

During the summer of 1858, while the gold-seeking parties from Georgia, the Cherokee nation, Missouri, and Kansas were on their prospecting tours, a number of ill-founded and exaggerated reports were published in the newspapers of the country. The most pertinent of these stories are given below. People gave these first reports the scant attention they deserved. Then, near the end of August, John Cantrell and other mountain traders arrived at the Missouri frontier with yellow dust in their pouches and with the first authentic story of gold discoveries. They reported that William Green Russell, and the twelve men who had stayed with him when all others turned back discouraged, had found good diggings on the South Platte near the mouth of Cherry creek. The sight of gold, even though in small quantities, caused a furore of excitement, and the contagion quickly spread throughout the country.

In early September parties of gold seekers began to form on the Missouri frontier and to set forth for the purported Eldorado. Throughout the autumn of 1858 other eager argonauts continued to push westward. Conflicting reports came back from the mountains.

Some returned gold seekers and most of the letter writers from the mines gave encouraging reports; good wages were being made and prospects were flattering. Others, discouraged gold seekers, said that the whole thing was a humbug, promoted by land speculators and town-lot owners. Those who took the middle ground were convinced that though gold had been found over a wide area it did not exist in sufficient quantities to pay for the work of gathering it. Many pointed to "color" in the sand and gravel east of the mountains and argued that this "float gold" had been washed down from the hills and that the ore bodies from which this "dust" had come would certainly be revealed by thorough prospecting.

Throughout the winter the majority of those who remained at the base of the Rockies were hopeful. They did very little mining and that little did not pay. Letter after letter says that the writer is busy putting up a house or hunting and has had no time to mine; or else the excuse is made that the streams are frozen and it is too cold to mine; but when spring comes big returns are to be expected.

Enthusiasm was thus maintained through the country, with the most favorable stories going the rounds of the press. As a result thousands of persons made preparations for a trek to the new gold country. Early in the spring the vanguard – footmen and handcart men – pushed across the plains. Early arrivals at the mouth of Cherry creek found no paying placers. Sorely disappointed, they turned about, denounced bitterly the "Pike's Peak humbug," and set their faces for home. Upon meeting the oncoming tide of emigrants, they turned most of them back, disillusioned, to the states.

It was late in the spring before any gold veins were found. The first, the Gregory lode (near present Central City), was discovered on May 6, 1859. Soon others were revealed. The saving discoveries were made just in time to prevent collapse of all hopes for the new country. The outlook now improved, paying mines were developed, and the permanent foundations of Colorado were laid.

The contemporary reports and letters reproduced here not only tell the general story of events, but they portray as well the feeling of the time and reflect the hopes and disappointments of the would-be miners and commonwealth founders. From these colorful and interesting records we gain a more adequate conception of the character and importance of this great stampede – one of the typical and outstanding gold rushes of Western history.

GOLD REPORTS AND
DEVELOPMENTS OF 1858

GOLD REPORTS AND
DEVELOPMENTS OF 1858

LATEST FROM UTAH AND THE PLAINS[1]

Oliver P. Goodwin,[2] arrived here on Wednesday last, [July 21] direct from Fort Bridger, in 21 days, with an express to Majors & Russell.

... On the head waters of the South fork of Platte, near Long's Peak, gold mines have been discovered and 500 persons are now working them. These mines are now yielding on an average $12 a day to each hand.

[1] *Kansas Weekly Herald* (Leavenworth), July 24, 1858.

[2] Goodwin's account of a prospecting trip he made in 1849 from Fort Laramie to the Spanish Peaks, was published in the *Hannibal Messenger* of March 22, 1859. See below, page 229 for one of his letters. At the time of his death the following obituary appeared in the *Sons of Colorado* (Denver, Colorado), II, no. 8, p. 27:

"On December 23 [1907] death closed the remarkable career of Oliver P. Goodwin, a veteran of the Mexican war, a soldier of many of the Indian wars, a filibuster in Central America under "Billy" Walker, a fur-trader, freighter across the plains, miner in the early days of California and one of the first to be identified with the stock industry of northern Colorado and Wyoming. Goodwin was 77 years old and until a month before his death was active mentally and physically and had a mind stored with a fund of information of the West of early days. Probably no man living this side of the Mississippi had a more varied life than Goodwin, who was for five years master of a government transport when supplies were sent from Fort Leavenworth to the forts of Colorado, Wyoming, and Utah in the days when Indians were most dangerous. Goodwin had headquarters in Pueblo when it was merely a fur-trading post and trekked all over this locality for the St. Louis Fur Company when there was not a cabin in the Cache la Poudre valley. He made and lost several fortunes before he was 40, and in the early '70s, having noted the value of northern Colorado for stock raising, bought land near Fort Collins and there conducted a big ranch until he went to Wyoming in the early '80s. Continuing in the cattle business, he again accumulated a fortune and two years ago went to Greeley. He was born in St. Louis, and was a Mason of high standing for half a century and is survived by a widow and daughter."

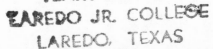

They are 175 miles from Fort Laramie, and 25 from St. Vrains Fort, in Nebraska.[3]

LATEST FROM FORT LARAMIE [4]

Major C. [A.] Dripps,[5] one of the oldest mountaineers now living, arrived home yesterday morning, having left Fort Laramie on the 4th of July, and passed thousands of wagons. The major says for the last twenty years he has never known the road to be in as bad traveling condition. All along the route teams were stuck in the mud. At the crossing of the South Platte they were using fifty yoke of oxen to cross one wagon. . . The major also reports that there have been discoveries of gold at Bent's old fort, and that over one hundred and fifty men are now at work in the diggings, making from eight to ten dollars per day. The gold discoveries are creating great excitement in the mountains.

LETTERS FROM KANSAS [6]

Topeka Kansas, August 7, 1858

TO THE EDITOR OF THE BOSTON JOURNAL: . . . several weeks since a party of young men from Lawrence, influenced by reports which have been coming from the west for several months past, started for the mountains in the western part of the territory, some six hundred

---

[3] On his way eastward Goodwin doubtless learned of the presence of gold seekers on the upper waters of the South Platte. On July 1 there were 104 men at the mouth of Cherry creek (site of Denver), but because they found little gold all but thirteen had turned back by July 6. (See the preceding volume in this *Southwest Historical Series*, 71, 111.) So Goodwin's report as to the number of men and their earnings was greatly exaggerated.

[4] Kansas City *Journal of Commerce*, July 30, 1858.

[5] Andrew Drips was famous in the fur-trade of the far west and as Indian agent. He was born in Pennsylvania in 1789 and died at Kansas City, September 1, 1860 (South Dakota Historical *Collections*, IX, 169).

[6] Published in the *Boston Journal*, August 16, 1858. An editorial in the same paper called attention to the letter from Kansas.

miles from this point, on a gold hunting expedition. A letter has just been received from one of the expedition, stating that they have met with more than hoped-for success, and that the hitherto unexplored region to which they have gone undoubtedly abounds in gold mines of the richest character. Further letters, containing detailed information, will probably be received within a few days. Meanwhile much interest is felt in the matter and if the intelligence is confirmed there will immediately be a great "rush" for the auriferous locality.

LETTER FROM KANSAS [7]

Lawrence, K.T., August 10, 1858

Reports of gold on our western border have made quite a stir in our community, reawakening the *auri sacra fames* of our California adventurers, some of whom have gone in pursuit of the glittering treasure. Their friends are apprehensive they will have a hard time and come back leaner than they went in body and estate.

GOLD AND PEARLS IN KANSAS [8]

We find the following in the *Grasshopper*, Aug. 14, and give it for what it is worth:

"Numerous companies are being formed in the territory for Pike's Peak, the gold region of Kansas. We understand that a letter received by a gentleman of Lawrence, from a son at the mountains, represents the prospects very favorably. Diggers are obtaining from $50 to $60 per day, and pieces of gold have been found, varying from $50 to $250 in value. Indians of various tribes, on the annual buffalo hunt, have forsaken their

7 *Boston Evening Transcript*, August 19, 1858.
8 *Leavenworth Times*, August 21, 1858.

chase and 'gone West,' in search of gold. Furthermore, it is stated to us on the most reliable authority, that in the White river, a tributary of the Kansas, pearls have been found, the largest of which are about the size of peas. Some of these have been sent East for the purpose of ascertaining their value. Meantime eager parties continue the exploration of the stream."

RICH DEPOSITS OF GOLD NEAR FORT LARAMIE [9]

The *St. Joseph Gazette* says: "We learn from S. Tennent, Esq., that gold has been found in large quantities about seventy miles from Fort Laramie, in the direction of Laramie's Peak. A young gentleman, Wm. Bryan, formerly of this city, has written a letter to his father, living in Kansas, urging him to leave everything here and go to those mines. He writes that he was shown one lump that weighed four pounds, and was assured by those who made the discovery that these mines equal the richest of California placers."

The fact that this region of country has been so frequently passed over by emigrant trains, and subjected to repeated mineralogical surveys, both private and governmental, favors the assumption that the whole story is apocryphal, and it would be folly in any one to go there in search of gold until the report receives further and abundant confirmation.

FROM FORT LARAMIE [10]

A private letter to a gentleman of this city, written from Fort Laramie, has been shown us. The date is a day or two later than that received through the St. Joseph papers, mentioning gold discoveries, and published yesterday; but the writer makes no mention

[9] *Daily Missouri Republican*, August 25, 1858. This story was reported in the *Boston Evening Transcript* of August 30, 1858.
[10] *Daily Missouri Republican*, August 26, 1858.

whatever of such rumors. The writer was in a position to give him early knowledge of the subject if any discovery of that kind had been made. Nothing new from the troops at the fort or in the vicinity.

[The *Omaha Times* of August 26, 1858, quotes the item from the *St. Joseph Gazette* and adds:]

We would inform the Gazette that gold quartz has been found upon the Platte bottoms to the westward of here some 30 miles, and reports assure us that some three or four dollars per day can be made washing from the bed of the same stream not over one hundred miles westward from here. That gold has been found upon the head waters of that stream we do not doubt and were we a prophet, or a son thereof, we would predict the day not far distant when the gold diggings of Nebraska would create as much excitement as those of California.

FROM THE PLAINS CORRESPONDENCE OF THE
REPUBLICAN [11]
Rulo, Nebraska ter., August 22 [1858]

Mr. Editor: Our fellow-citizens, Charles Martin and Wm. Renceleur [Kenceleur],[12] have just arrived from the Platte bridge. They made the trip to this place in seventeen days. Their partner in the bridge, John Richards,[13] Esq., came with them. The news they bring is cheering. The plains are alive with men, teams and business. Gold has been discovered along the South Platte, on Cherry creek, and they bring with them a

---

11 This appeared in the *Daily Missouri Republican* of September 1, 1858.

12 These men are identified as early settlers of Rulo (in the extreme southeast corner of Nebraska) in Nebraska State Historical Society *Publications*, XX, 309.

13 John Richard [sometimes rendered Reshaw or Richaud] was a prominent trader of the Fort Laramie region during the 1840s and '50s. He set up a trading-post in pioneer Denver.

specimen of the dust, which is very beautiful. I have in my possession a small portion of the specimen brought by them. A company is about organizing to start from here immediately to the mines, and several other companies will leave here early in the spring. They state that a man can work out from ten to fifteen dollars per day in a common pan.

The place where the miners are at work is not over about five hundred miles from this place, and the above named gentlemen state that the country is very fertile, being just at the foot of the mountains, and that the timber is abundant of every kind. They also state that fair crops of corn have been raised on the Platte this season in many places; among others, near Fort Laramie and at the Platte bridge. You will hear of this matter, however, from other sources in a few days as Mr. Richards leaves here to-day, on the packet, for your city, to purchase goods for the Platte. Yours truly, M. H. W.

[FIRST IMPORTANT AUTHENTIC PUBLISHED ACCOUNT OF THE GOLD DISCOVERY] [14]

THE NEW ELDORADO!!!/ Gold in Kansas Terri-

[14] Kansas City *Journal of Commerce*, August 26, 1858.

This story, in abbreviated form, appeared in the *Sunday Morning Republican* (St. Louis) on August 29, as a telegraphic dispatch dated at Kansas City, August 26. It appears in the *Boston Daily Journal* of August 30, and in other papers of the country. The *Journal of Commerce* was generally credited with publishing the first report of the gold discovery, but this was disputed by others. Witness this item in the *Leavenworth Times* of April 2, 1859: "The *Journal of Commerce*, published in Kansas City, Mo., claims to have promulgated the first announcement ever made by a newspaper of the discovery of gold in the Pike's Peak region—on the 26th of August, 1858. This is but a sample of the usual mendacity manifested by the editors of that sheet. Information of the existence of gold in that quarter was made public in this city and was telegraphed all over the country prior to the date mentioned by the *Journal*."

The *Journal of Commerce* story was copied in the *Lawrence Republican* of September 2, 1858.

tory!!/ The Pike's Mines!/ First Arrival of Gold Dust at Kansas City!!!

We were surprised this morning to meet Mons. Bordeau [15] and company, old mountain traders just in from Pike's Peak.

They came in for outfits, tools, etc., for working the newly discovered mines on Cherry creek, a tributary of the South Platte.

They bring several ounces of gold, dug up by the trappers of that region, which, in fineness, equals the choicest of California specimens.

Mr. John Cantrell, an old citizen of Westport, has three ounces which he dug with a hatchet in Cherry creek and washed out with a frying pan.[16]

Mons. Richard, an old French trapper, has several ounces of the precious dust, which he dug with an axe.

Mons. Boesinette [17] has several rich specimens.

The party consists of nine men, all of them old mountaineers, who have spent their lives in the mountains. Mons. Bordeau has not been in the states for nine years, until the present time.

We have refrained from giving too great credence to these gold discoveries until assured of their truth, but it would be unjust to the country longer to withhold the facts of which there can no longer be a doubt.

Kansas City is alive with excitement, and parties are already preparing for the diggings.

[15] James Bordeau was in charge of Fort Laramie in 1842 and for several years thereafter. See Hafen and Young, *Fort Laramie,* 83, 116, 121. In the middle '6os he had a ranch about ten miles below Fort Laramie, where he lived with his Indian wife and family.—E. F. Ware, *The Indian War of 1864,* 271.

[16] For an account of Cantrell's visit to the mines, see pages 71-72 in the preceding volume of this series.

[17] Bissonet is frequently encountered in the Fort Laramie region in the 1840s and '50s. For biographical data on Louis and Joseph Bissonet, see J. C. Luttig, *Journal of a Fur-trading Expedition,* edited by Stella M. Drumm, 148-149.

In order to give a correct idea of the locality of these new mines, we will state that they are on Cherry creek, one of the most southern branches of the South Platte, in the center of the best hunting grounds of the Rocky mountains. Game exists in great abundance, and plenty of timber, water and grass. They are in latitude 39 degrees, and doubtless extend to all the streams of that region. The waters of the Arkansas and the South fork of the Platte rise together about the same parallel, and no doubt all partake of the same auriferous character.

For the Gold Regions [18]

There is perfect furore in Kansas City on the subject of the Pike's Peak mines. We have no hesitation in saying that in a fortnight the road from Kansas City to Council Grove, and thence west to the mines, will be lined with gold seekers for the new Eldorado. This is the best route as there is a splendid road for two-thirds of the distance, good stopping places; plenty of feed for cattle and men for the first two hundred miles, and shorter by several days than by any other route.

Another Gold Hunter

We met yesterday, Mons. Charles Primo, one of the agents of the American Fur Company, and one of the best business men in the mountains. He, too, is in for an outfit for the new gold diggings. He says that men have dug out over $600 per week, with nothing but their knives, tomahawks and frying pans. He will be in Kansas City ready to start in a short time.

Already has the gold fever excitement created a great demand for mules in our market. It is generally believed that mule trains will make the quickest and most successful trips.

---

[18] Kansas City *Journal of Commerce,* August 27, 1858.

GOLD EXCITEMENT ON THE INCREASE [19]

Dr. R. R. Hall is now making preparations to leave at an early day for the new discovered gold regions near Pike's Peak. His route will be the old Santa Fe road to the crossing of Sand creek, at which point he is only a short distance from the mines. He contemplates leaving with five or seven wagons, taking out a stock of groceries and provisions, and if immediate application is made the doctor will make arrangements to take from ten to fifteen passengers, furnishing them with provisions on the route.

Great excitement is now rife in this city, and on every corner can be heard the usual conversation incident to the gold fever mania. "How far is it there?" "How long will it take to make the trip?" "Do you think the information is correct?" "Have you seen any of the dust?" "What kind of an outfit will it take?" etc. etc. Such are some of the thousand interrogations that we now hear made in every quarter of our city.

All persons wishing to make arrangements to emigrate, will do well to call upon Dr. Hall and learn from him the full particulars.

NEW GOLD DISCOVERY [20]
(From the *Westport Star*, Aug. 28)

Mr. John Cantrell, well known in and a long resident of Westport, has just arrived, being 21 days out from Cherry creek mines. This is a "sand creek," and flows into the South Platte. He brings specimens of gold that he dug himself. It is what is known as "float gold," and having seen it, we unhesitatingly pronounce it the pure stuff.

[19] Kansas City *Journal of Commerce,* August 27, 1858.

[20] *Marshall Democrat,* September 3, 1858. This story was also copied in the *Lawrence Republican* of September 3, 1858.

Mr. Cantrell being at Fort Laramie, heard of a prospecting party on the South Platte, and immediately started. He met a party of 15 [13], headed by Green Russell, of Georgia, from whom he learned that they had been successful. Seven of them had made over $1000 in ten days. Examining the places visited by Russell's party, about 4 miles above the mouth of Cherry creek, Mr. Cantrell found that the dirt would yield from 17 to 20 cents to each pan; and he thinks that if properly worked, one man can make from $20 to $25 per day. The mines will average with those of California, in which Mr. Cantrell is experienced, having spent several years in them.

The country in the neighborhood is healthy, it is good for farming; grass and game abundant. Mr. Cantrell has two or three bushels of sand from the mines on the way in, which will arrive in a few days, and persons who are curious, can examine for themselves.

Mr. Cantrell is the only person direct from the mines. Being well known here, his word is undoubted, and we do not hesitate to let the world know his opinion and hear the news he brings, feeling confident that his statements may be relied upon.

Arrival of John Cantrell [21]

Mr. John Cantrell, of Westport, arrived home on Thursday, from the gold mines at Cherry creek and Pike's Peak. Mr. Cantrell is well known to the people of this region of country, and his statements are of great interest. He left the mines on the 20th of July, at which time there were from thirty to thirty-five people at work, digging and prospecting.

[21] Kansas City *Journal of Commerce*, August 28, 1858.

Mr. Cantrell and party prospected all the small streams in the vicinity of the first discovery, and found gold upon all of them. They also found it from two to eight feet below the surface. This corresponds exactly with the first discoveries of gold in California.

Mr. Cantrell has with him samples of the gold, washed, in the sand, and in the quartz.

The washed gold is the regular placer gold, in fine scales of extraordinary purity and richness. This was scraped up at various points during his prospecting tour.

He has also samples of the gold mixed with black sand, taken from the banks of the streams and from the bedrock. To an old miner this black sand settles the question, even without the gold itself, as it is infallible index to rich diggings. Mr. Cantrell also has a few specimens of quartz picked up away from the streams which contain gold, but from the want of proper tools and other facilities, no examination could satisfactorily be made in this respect.

A portion of Mr. C's party remained at the mines, and will continue their prospecting operations until the return of the remainder and Mr. C. with tools and outfit. . .

Many of our old mountaineers, and young men, intend starting immediately, with the design of prospecting as much as possible the present season; but we do not think that a large emigration would be able to get through the winter without great hardships, and for those who have the patience to wait, and who design staying in the country, the first of March is the proper time.

JOHN CANTRELL [22]

We had a call from Mr. Cantrell, on yesterday, with whom we had a lengthy and satisfactory conversation about the new mines of Pike's Peak.

Mr. Cantrell is an old California miner himself, and has had much experience in gold digging. He went out last spring with a lot of goods to trade with the army and emigration crossing the plains. After closing out his stock, and hearing that several parties had been prospecting on the headwaters of the Platte, he determined to visit the country and see for himself.

He first arrived at Cherry creek, and there learned that a party of about one hundred men, a portion from Arkansas and Missouri, and a portion of Cherokee Indians, had been in the country for several weeks during the spring, but from the unusually high waters, the country being flooded beyond what had ever been known, they became discouraged, and many of them returned home. He learned, however, that a party of fifteen [13] men, under Green Russell, of Georgia, had remained and were up nearer the mountains. He set out to find this party, which he succeeded in doing. They had been at work for ten or twelve days and had taken out about $1,000, a portion only being at work, and with no mining tools.

Mr. Cantrell traversed about seventy miles of the country, and in every stream he prospected he found gold. In one place where he prospected, three pans of dirt paid 66 cents, weighed with apothecaries scales.

Mr. Cantrell brought about three bushels of dirt in his wagon, which was tested in Westport yesterday

[22] Kansas City *Journal of Commerce,* September 1, 1858. The *Herald of Freedom* of September 11, 1858, copies a story from the Kansas City *Journal of Commerce* of August 29, telling of the arrival of John Richard and his report of gold discoveries.

morning, by Edward Payton and other experienced miners, and which yielded about fifteen cents to each two quarts of dirt, and Mr. C. informs that he prospected bars that would yield from 20 to 25 cts to the pan.

BY TELEGRAPH [23]
Leavenworth, August 29, per U.S. Express
Company to Boonville, August 30

Considerable excitement exists in Lawrence and Kansas City in consequence of recent arrivals from the gold region of Pike's Peak, confirming the existence of the ore in abundance at that locality. A company which went from Lawrence in June had met with good success. The gold found at these diggins is similar to that of Fraser river and California.

A Mr. Richards arrived at Kansas City on the 28th, reported that with very limited prospecting satisfactory results were obtained. Two men with inferior implements, washed out $600 in one week, on a small stream fifty miles from Pike's Peak.

A second Fraser river is apprehended.

THE GOLD FEVER IN WESTPORT [24]

We are informed that on Sunday last [August 29], there were "signs" of emigration to the Pike's Peak gold region seen in the streets of Westport. A wagon loaded with the necessary outfits for mining, consisting in part of wheelbarrows, plows, washers, pans, picks and whiskey, was paraded in front of the Smith House, early in the morning, destined for Pike's Peak, D. M. Ross, agent – so says our informant. There will be more "signs" in Westport before the snows of winter.

23 Published in the *Daily Missouri Republican*, August 31, 1858. The same story appeared in the *Boston Evening Transcript*, September 1, 1858.
24 Kansas City *Journal of Commerce*, August 31, 1858.

MORE ABOUT THE GOLD DISCOVERIES ON THE PLATTE [25]

A gentleman of this city has put us in possession of a
letter dated at Fort Laramie on the 18th August, giving some additional and more reliable particulars in
regard to recent discoveries of gold in the South Platte
river. We attach more importance to the statements
of this because the gentleman who received it here
vouches for the entire truthfulness of the writer. We
give the pith of this letter.

At the time of writing he had just returned from
Cherry creek. The Cherokee company had returned
home before the writer arrived there, having prospected Cherry creek, Ralston's fork and Long's creek,
without having found much gold – thinking it would
not pay, they became discouraged and went home.
Capt. Russell, of Lumpkin county, Georgia, who was
with them, remained to prospect still further, and after
their departure was very successful. The writer saw
where they had been digging, and from the amount
of work done at each place, and the amount of gold
obtained, he thinks the prospect a very good one. The
product is said to be very irregular. The first diggings
are about four miles up the Platte river, and about
half a mile from the river towards Cherry creek. Here
two or three men would work with a rocker while the
others were on the look-out for better diggings, and
they made from seven to ten dollars per man each day.
After working here a few days, getting all the gold
they could, they moved out about three miles, and
rather up the river in a ravine: here they worked in
the same way, making from eight to ten dollars per
day, till the diggings failed. Then they moved again

<hr>

[25] *Daily Missouri Republican*, September 1, 1858. The substance of this
letter was copied in the Kansas City *Journal of Commerce* of September 5,
1858.

to the river about six miles from the first work, and
the proceeds were about as good. Mr. Russell says he
has gotten as much as three dwt. per pan, and the three
men with the rocker have obtained from one day's
washing of one hundred buckets of earth, which they
had to pack fifty yards, 49 dwts. and 2 gr. As a result
of all their work and prospecting they obtained something this side of five hundred dwts. of gold, or about
twenty-five ozs. . .

New Gold Discoveries [26]

Oh the Gold! the Gold!—they say
'Tis brighter than the day,
And now 'tis mine, I'm bound to shine,
And drive dull care away.

The following article clipped from the Westport
Star of Empire, and Kansas City *Journal of Commerce,* seems to fully corroborate the news of the discovery of gold on Cherry creek, which we published
last week. Quite an excitement has been created in this
city, and there is strong talk of the immediate organization of another party to start immediately for the
diggings. The original party outfitted here, and other
parties will find this the most favorable point for obtaining pans, rockers, and all other necessary articles
for the expedition. . .

From Kansas [27]

Leavenworth, Kansas, 2d.

The Pike's Peak gold excitement is on the increase.
Two old Californians came in yesterday to make arrangements for working the mines successfully. One

---

[26] *Lawrence Republican,* September 2, 1858.
[27] *Boston Evening Transcript,* September 6, 1858.

company left for the gold region yesterday, and others
are now organizing.

Gold News [28]

It is an old saying, "that it never rains but it pours."
Thus, just in the midst of the excitement about the
Pike's Peak mines, we have a gentleman arrive from
the head waters of the Arkansas, still south of the Peak,
with samples of gold from one of the tributaries of
the Arkansas, of a better variety of diggings than any
we have yet seen. They embrace the lump or "nugget"
gold, as well as the float gold from the beds of the
streams. The gentleman is from Illinois, and had left
town before we had an opportunity to see him.

The specimens can be seen at the jewelry store of
Jos. Gilliford, on Main street. The locality where it
was found is represented to be of unusual richness, and
well adapted for working.

Ho! for the Mines

A company of sixty men are now preparing to leave
this city for the gold regions. They will leave as soon
as their mules, wagons, and outfits can be got ready,
which they expect will be inside of ten days. Another
company is forming under charge of Dr. R. R. Hall,
who also expect to be off by the middle of September.

There is a friend of ours, a surveyor by profession,
who is preparing to leave with another company, and
to show what young America can do when aroused —
he is provided with picks, shovels, pans, camp kettles,
and all the et ceteras necessary for mining life. He
has also a compass, transit, level, chain, drafting in-
struments, paper, and all the other requisites for surveying. His plan of operation is to prospect the country with his pick and pan, until he is satisfied where

---

[28] Kansas City *Journal of Commerce*, September 3, 1858.

to squat, when he will proceed to locate, preempt, survey, plat, map and lay out a town, and go into that line of business on his own hook. Such are the men who are now leaving Kansas City for the gold region. It will not take such boys long to develop the country. Hurrah for young America.

KANSAS GOLD FEVER [29]
(Correspondence of Mo. Democrat.)

Kansas City, Mo. Sept. 3, 1858

The excitement caused by the discovery of the Pike's Peak gold mines is still unabated, indeed, it is vastly on the increase. Every man has gold on his tongue, if none in his pocket. The first question in the morning, after coming down town, is, "What's the news from the gold diggings?" Old men, young men, women and children, may be seen in groups, discoursing the merits and demerits of the wondrous discoveries of gold "almost in our midst." If there is not an abatement of this feeling before spring, our city will be depopulated. . .

Col. Bent,[30] of Bent's Fort, is loading his teams in this place now for the purpose of exporting goods to the mines. He expects to establish a trading post there, and will transport part of his supplies this fall.

ASBURY

LETTER FROM KANSAS [31]
(From our regular Kansas correspondent.)

Pardee, Atchison co., K.T., Saturday, Sept. 3, 1858

TO THE EDITOR OF THE BOSTON JOURNAL: . . . As you have perhaps learned by some of the river papers, the latest news from the gold mines in western Kansas

---

[29] *Missouri Democrat,* September 8, 1858.

[30] William Bent was proprietor of Bent's Fort, most famous fur-trading post of the Southwest.

[31] *Boston Daily Journal,* September 14, 1858.

is of the most encouraging character. A gentleman who arrived in St. Joseph from Laramie's Peak, a day or two since, reports that the miners there are realizing from ten to twenty dollars per day; and a party of old mountaineers recently arrived in Kansas City, bringing with them specimens of gold of a very fine quality, which they had found in abundance among the head waters of the Arkansas. They procured the necessary tools and outfit for miners, and immediately returned. There is a good deal of excitement in Kansas City, and a party is nearly organized to leave for the "diggings." A. D. R. [Albert D. Richardson] [32]

Gold! Gold!! Gold!!! Gold!!!!
Hard to Get and Heavy to
Hold.
Come to Kansas!!
California and Frasers river "no whar!" . . .
Cherry Creek and Pike's Peak Ahead!!!
Great Excitement!! The Atlantic Cable
not Thought of!!!
[Heading of *Kansas Weekly Press* of Elwood, September 4, 1858. Particulars follow. They feel sure of a heavy emigration]

[32] Albert Deane Richardson, prominent journalist of the period, was to visit the mines in 1859 and with two other newspaper men—Horace Greeley and Henry Villard—issue a famous statement about the mines. We shall hear much from him hereafter.

Richardson was born in Franklin, Massachusetts, October 6, 1833. After some experience in the newspaper field, he went to Kansas in 1857, and from there wrote graphic accounts of slavery troubles. His experiences in the west are entertainingly told in his *Beyond the Mississippi* (1866). He followed the Union armies as a correspondent, was captured and for nearly two years was held a prisoner. According to *The Trail* (Denver), v, no. 3, p. 8, he was killed in New York City by a man named McFarland, December 2, 1869. For biographical sketches see *Kansas Magazine*, I, no. 1, pp. 15-18; and *National Cyclopedia of American Biography*, VIII, 465.

A Busy Day – First sale of goods for Mines [33]

Yesterday was a busy day with our merchants on the levee. Trade was made with buyers from all portions of the interior and the country south from the mines of the Rocky mountains, and the pueblos of New Mexico, to the valleys of the Osage and the surroundings of the Oxford precinct.

The first sale for the gold regions of the Arkansas, and the southern tributaries of the Platte, was also made yesterday.

Mr. J. T. Jones bought of Messrs. J. S. Chick & Co., a bill of goods, comprising every thing that is needed to constitute a stock in trade of a miner's merchant. His purchases amounted to nearly a thousand dollars. He is to start his teams immediately, intending to winter either in the mines or some neighborhood near at hand, where he will be able to run his goods into the gold region market early in the spring.

Nebraska Correspondence [34]

Nebraska City, September 7th.

The "gold fever" is raging, and many are preparing to go to "Cherry creek" instantly. The most flattering reports are reaching us daily, in regard to the immense success at the "diggins," but I have serious doubts of their truth.

Wm. Bent's Train [35]

Some forty odd Frenchmen arrived here Sunday morning, [September 5], by the *War Eagle,* to go out with Col. Wm. Bent, to his fort upon the Arkansas, near the new gold regions, where they are to be em-

---

33 Kansas City *Journal of Commerce,* September 7, 1858.

34 *Daily Missouri Republican,* September 17, 1858.

35 Kansas City *Journal of Commerce,* September 7, 1858.

ployed by the colonel, about his fort, trains, trading post, and the mines.

OFF FOR THE MINES [36]

Mr. S. H. Clark, agent for the United States Express Company, received yesterday morning by the steamer *A. B. Chambers,* a letter from Jefferson City, instructing him to secure a passage for six young men, in the Pike's Peak mining company that leaves that city on the 15th inst., in company with Col. Wm. Bent's train for his fort in the mountains, which is some forty or fifty miles distant from the gold discoveries on Cherry creek.

Including the Jefferson City party, there will probably be from twenty to twenty-five in the company, that will leave under the escort of Col. Bent and his train; other parties, we understand, are making arrangements to leave at an early date.

MAJOR DRIPP'S TRAIN [37]

This veteran trader will leave with his train for his trading post near Fort Laramie, tomorrow. He takes out five heavily laden wagons, freighted with Indian goods. There is no man connected with mountain commerce, who enjoys a higher character than Major Dripps. For over thirty years he has been employed in trade with the Indians, and from the Yellowstone to St. Louis, his word is a bond for all who know him. He informs us that as soon as he arrives at his trading post, he intends to take a trip to the new gold mines, and establish a trading house in that region. We expect to hear from him about the first of January, fully in regard to the new diggings. His many friends wish

[36] *Ibid.,* September 7, 1858.
[37] *Ibid.,* September 7, 1858. This was Andrew Drips, fur-trader.

him a successful trip to the mountains, and a prosperous season of business.

GOLD EXCITEMENT [38]

From private letters received in this city, there remains but little doubt of the correctness of the reports relative to the new gold discoveries on Cherry creek, a tributary of the South Platte-Nebraska territory. Already are companies preparing to set out for the new diggings, and if no adverse accounts are received there will be about 25 of our citizens enroute within the next two weeks. Some are almost wild with excitement, and wish to start immediately, but the older and wiser portion advise them to hold on for a few days until something more definite is ascertained.

Several who intend to go are old California miners, who understand the business, and if there is gold there, we shall expect to see the Council Bluffs boys bring home their share of the rocks.

THE GOLD HUNTERS [39]

It will be seen by our advertising columns that the gold crusade has already commenced. Some three or four parties will leave Kansas City within the next fortnight for Pike's Peak and the head waters of the Arkansas and South Platte. These parties are mostly young men, many of them inured to life in California and on the western plains. They are not men who go without knowing what they are to encounter, or with a vague idea of what is before them. They will be well provided with all equipments necessary to pass the winter in an uninhabited country, and expecting to pass such a winter.

[38] *Council Bluffs Bugle,* September 8, 1858.
[39] Kansas City *Journal of Commerce,* September 9, 1858.

More Gold [40]

Gold has been discovered in the western part of this territory, at Pike's Peak, on the Rocky mountains, near the headwaters of the Platte and the Arkansas rivers. As usual, it has caused great excitement throughout the country, and parties are starting from various points, for the new diggings. Many persons will rake and scrape up all the money they can gather, and proceed to the gold regions, where they will probably meet only with disappointment, spend their means, and be left destitute; while with that money they could have wintered comfortably at home, and found themselves healthy and vigorous in the spring. Where one makes a fortune in the gold diggings, ten come away worse off than when they went.

Gold – Pike's Peak [41]

Just now along the river at every town the great subject conversed upon is "Pike's Peak, Pike's Peak!" Here the rejoicing over the postponement of the land sales has taken its place for a few days principally, yet you hear "Pike's Peak" at almost every corner you pass.

There is yet no intelligence sufficiently reliable, we think, to warrant a stampede to "Pike's Peak." . .

We advise those taken with the "Pike's Peak" fever to not overdo themselves; we think the disease is not dangerous, and will pass off without any serious results, by taking a slight dose of reflection.

Gold in Nebraska [42]

Rich diggings discovered within twelve days travel from Omaha City – Great excitement on the plains – The mountaineers flocking to the mines – Omaha City

40 *Kansas Chief* (White Cloud, Kansas), September 9, 1858.
41 *Nebraska Advertiser* (Brownville), September 9, 1858.
42 *Omaha Times,* September 9, 1858.

merchants preparing to outfit miners – Miners making from $10 to $15 per day with pans, etc.

On Saturday last we spent a few hours with Messrs. E. S. & E. A. Owens, who have just returned from Salt Lake, and they not only confirm the report of rich gold diggings in Nebraska, about 450 miles travel to the westward of this place, on Cherry creek, but found the excitement prevalent and intense from beyond Ft. Laramie to Ft. Kearny and nearly all the mountaineers hurrying with their families and stock to the diggings. All along the route men hastening on to these mines, were met, and although not miners – these now numbering about three hundred – were using pans entirely in obtaining gold, they were making from $8 to $15 per day per man, and as soon as rockers could be made and used, the yield would not fall short of $100 per day per man. . .

A few weeks more and this fall's emigration will doubtless be hurrying on through here to the mountains. They are near enough to the settlements and military posts to feel under no apprehension from famine or the Indians, and in a country rich in agricultural resources. . .

PIKE'S PEAK – GOLD EXCITEMENT IN THE CITY – A NEW CALIFORNIA [43]

Leavenworth was agog Thursday. There was a buzz of excitement throughout the entire day. "Gold" was upon the lips of all, and "Pike's Peak" acquired an immense deal of notoriety. The excitement of election was forgotten in the eager anxiety to learn something concerning the new El Dorado.

Mr. Elmore King [44] gave us a brief account of his

---

43 *Leavenworth Times,* September 11, 1858.

44 King went out with the Cherokee party in the spring. Much controversy ensued as to the truthfulness of his report.

\

experience in the gold region. He went out in the spring and left on the 27th of last July. During this time he was engaged in prospecting, and never failed to discover the "pure ore." Several companies were regularly engaged in the work of .digging, and, as a general thing, took out from $5 to $10 a day per man.

All they had was common pans, and even these were scarce. A man, with proper tools for obtaining gold, could secure from ten to fifty dollars a day. . .

In addition to the information received from Mr. King, we obtained much important intelligence from an old mountaineer, by the name of Hyatt, who has lived in the vicinity of Pike's Peak for the last ten years. He corroborates all that Mr. King asserts, and says that "California can't hold a candle to this new gold region." . .

We have hitherto refrained from endorsing the rumors and vague reports that have come to us in regard to Pike's Peak, for we did not desire to have our readers deceived or to encourage any false excitement. But we now have stable facts to go upon, and do not hesitate to say that this newly opened country will prove another California in wealth. There is no mistake about the gold. It is here to be seen and examined. It speaks for itself. Corroborative and conclusive statements establish the truth, and it is safe to say that the right sort of men will soon have an opportunity of realizing their eternal fortunes.

GOLD EXCITEMENT – PIKE'S PEAK![45]

The Pike's Peak "mad" is raging fiercely in this

---

[45] *Leavenworth Journal,* September 11, 1858. Another story in the same paper tells of the return of Elmore Y. King and his report of big earnings in the gold mines. The *Leavenworth Weekly Ledger* of September 12, also tells of King's return and presents the advantages of Leavenworth as an outfitting point.

community, and threatens to "carry off" a number of our denizens. The fever will continue to increase for some days, when it will abate, like another "yellow" fever, upon the approach of frost. As an evidence of the malignity with which the epidemic spreads, meet a hitherto quiet and staid delinquent and present his bill; he will answer "Pike's Peak!" Drop into a barber shop to undergo the simple act of shaving or having your hair cut, and you are met with the interrogatory: "Have it done in Pike's Peak style, sah?" Sit down to dinner, call for dessert, and the waiter wants to know if you wish your beef cooked "a la mode Pike's Peak," and your pudding with "Pike's Peak sauce."

Quite a number seem to have resolved upon trying their luck in that land of gold, and no doubt dream of the sudden acquisition of huge piles of "Benton mint drops," dug already coined. We are disposed rather to dissuade than encourage every one who designs visiting that region this fall. . .

GOLD! GOLD! GOLD! [46]

Reliable reports concerning the recent gold discoveries in the vicinity of Pike's Peak still continue to arrive. Every trader or prospector coming from the region gives flattering accounts. The gold has been discovered in all the streams flowing from the mountains, but owing to the want of good tools for working, it has not been obtained in large quantities. Eight, ten, and fifteen dollars per day have been obtained with the simplest tools; and those who ought to know, say that with the proper tools fifty dollars can be obtained per day.

----

46 *Kansas Weekly Herald*, September 11, 1858. A. D. Richardson's letter of this same date, written from Sumner, Kansas, and published in the *Boston Daily Journal*, September 20, 1858, tells of the gold excitement and of the advantages of Sumner as a setting-out place.

The excitement is still on the increase, and spreading in every direction. St. Joseph, Council Bluffs, Nebraska and Omaha Cities; Leavenworth, Kansas City, Independence, Westport, and Lawrence, are all preparing to send forward a living stream of emigration to the gold regions. . .

GERMAN COMPANY FOR THE MINES [47]

We saw early yesterday morning the business committee of a company of Germans, engaged in buying cattle, wagons, chains, yokes, and other articles necessary to constitute the transporting outfit for their Pike's Peak mining company.

The committee was composed of Mr. Henry Morat,[48] Philip Swiekirt, and Henry Lubbe.

They contemplate leaving some time next week, and on their arrival, will go to work at once upon the pioneer arrangements necessary for a German settlement.

This is what we like to see – a German emigration among the first settlers of a new country.

They are the men who have been the founders or first builders of all the principal commercial cities of the union – New York, Philadelphia, Cincinnati, Chicago, and St. Louis. They have done as much for this city, as any other class of people. We wish them every success among the placers of the Rocky mountains.

INTERVIEW WITH MR. KING [49]

We had the pleasure of meeting in our office yes-

---

[47] Kansas City *Journal of Commerce,* September 11, 1858.

[48] "Count" Henri Murat and his wife, the "Countess," became notable characters in pioneer Denver. The count became a barber (he shaved Horace Greeley and charged him one dollar for the service). Countess Murat was one of the first women pioneers of Denver. See Louie Croft Boyd, "Katrina Wolf Murat, the Pioneer," in *Colorado Magazine,* XVI, 180-185.

[49] Kansas City *Journal of Commerce,* September 12, 1858.

terday with Mr. King, of Dayton, Ohio. He is just in from Pike's Peak, and brings the latest and most reliable news since the return of Cantrell, Bordeau and Richard.

Mr. King went out with a company from the Cherokee nation, in May last. They first commenced prospecting on the Arkansas, above Fort Atkinson. At intervals, while camping on that river, they tried the bed of the stream, and invariably found the "color," as far up as the old Fort Puebla, which is at the foot of the mountains. From this point the company struck across the country to Pike's Peak, and from thence to Cherry creek and Long's creek. There were but fifteen men at work when Mr. King left. They had no tools except their pans, and were making, on an average, $10 per day each man. . .

BY TELEGRAPH [50]

Leavenworth, September 12

Per U.S. Express to Boonville, September 15.

The Pike's Peak gold excitement is a perfect mania at this place. Two companies left for the gold region last week, and another will go this week. A large and well organized company, with Gen. Larimer [51] and Judge Hemmingway, bankers, at the head, will leave about the 25th inst., with six months' provisions and every thing necessary for wintering in the mountains, and mining in the spring.

The reports from the diggings continue flattering. There were several fine specimens of gold, brought in last week, now on exhibition here.

---

50 Published in the *Daily Missouri Republican,* September 16, 1858.

51 William Larimer was to become a prominent pioneer of Denver. A principal street and a Colorado county were named for him. Some of his letters will be given below.

LETTER FROM KANSAS [52]
Leavenworth, Kansas, Monday, Sept. 13, 1858
TO THE EDITOR OF THE BOSTON JOURNAL:

I promised in my last letter to write something further in regard to the gold region. The excitement which I predicted more than a month ago is now at its height. In fact the news of adjacent gold mines, where with the proper machinery men can make from ten to fifty dollars a day, ought to cause a stampede from a country where money brings five per cent, a month, and is difficult to obtain at that. It stirs men's hearts, as Theodore Parker says, "like the news of a line of gas pockets to the moon." Politics are forgotten. Speculation is ignored, and the latest news from Pike's Peak is the universal theme of conversation.

The intelligence is well authenticated, from several distinct sources. At St. Joseph a letter has just been received from an old resident of that city, who sends on for the proper tools, outfit, etc., and states that he is making ten dollars per day without them. In this city Mr. Elmore King, who arrived here last Thursday from the mines, having spent the season there, has been and is the hero of the day. He is a reliable gentleman, a native of Missouri a few miles from here. He will return to the mines in the spring with the proper outfits; reports gold very plentiful; says men with only a common pan are realizing from five to ten dollars per day, and with the proper tools could make from ten to fifty. In Lawrence several members of the party which went from there last spring, and have just returned for tools, outfits, etc., bring a similar report. With washers constructed from hollow trees, miners are making from eight to ten dollars per day. From Kansas City below, and several important river towns

---

[52] *Boston Daily Journal,* September 21, 1858.

above, in Nebraska and Iowa, we have similar intelligence from parties who have just returned from the mines.

Preparations are being made, all along the river, for starting for the gold region. . . A. D. R. [Richardson].

GONE TO PIKE'S PEAK [53]

During yesterday there was but little business done upon our levee save the outfitting trade with the gold hunters.

The town was fairly alive with the excitement occasioned by the bustle, trade, teams, and parting congratulations given to these fearless adventurers. We hope that every one of them will return at an early day, with several "half-barrels of the ore."

GOLD HUNTERS FROM SOUTH-WEST MISSOURI [54]

A company of five men arrived here yesterday from the extreme south-west counties of the state, for the purpose of buying an outfit for the Pike's Peak gold mines. One of the party informs us that hundreds will leave that portion of the state early in the spring for the Arkansas gold regions.

PIKE'S PEAK GOLD [55]

There is to be seen at the S[t]ar Saloon, below the post office, a large specimen of goldbearing quartz directly from Pike's Peak mines. The curious and the skeptical have an opportunity to see and be convinced. It seems to be very rich, and the sight of it is not calculated to allay the "gold fever." Go and see it.

[53] Kansas City *Journal of Commerce,* September 14, 1858.
[54] *Ibid.*
[55] *Leavenworth Journal,* September 14, 1858.

THE GOLD MINES! [56]

Pike's Peak meeting! Enthusiastic orators! The press and the "stump."

A large number came together last night, on the corner of Delaware and Second streets, to talk over the various matters connected with the gold mines, and to compare notes. We understand that the meeting was convened for the purpose of inducing emigration, or rather of presenting to emigrants the undeniable advantages which Leavenworth City possesses over all other towns on the Missouri river as an outfitting point. . .

Mayor Denman presided, and there were a number of inspiring speeches made by various gentlemen. It was clearly demonstrated that Leavenworth City could furnish outfits and transportation, at a moment's notice for ten thousand emigrants and that one firm alone had four hundred wagons & teams which can be had whenever required. Mr. Russell [57] it is said is prepared to fit out and land safely at Cherry creek two thousand men.

A large number are outfitting here, and will leave in a few days.

The most reliable reports substantiate the existence of gold in large quantities in the region of Pike's Peak, and the natural consequence is that there will be an immense rush to that region. Their own interest will prompt them to seek the best starting place, and as there is no rival town to divide the trade with Leaven-

---

[56] *Ibid.*

[57] W. H. Russell, of the great freighting firm of Russell, Majors, and Waddell. This company also maintained a large outfitting store at Leavenworth. The *Leavenworth Weekly Ledger* of September 12, 1858, carried a one-column advertisement of this firm, announcing goods and equipment suitable for the gold seeker.

worth City, it is but natural to conclude that Leaven-worth will soon be filled with gold hunters. Let them come.

GOLD WITHIN OUR REACH [58]

Hundreds flocking to the mines! Omaha City out-fitting the miners. The excitement in other places.

Since the organization of our territory, we have never seen an excitement as general and intense as it now is, concerning the newly discovered mines of gold to the westward of this place, some 400 miles. It is talked of on the corner of streets – in the streets – hotels, stores and everywhere. Men are preparing their camp outfits, some for a few weeks, and some for a winter's stay in the diggings. Merchants are busy, and "only just enough of the needful to purchase a camp supply," is eagerly sought after.

The *Elwood* (Kansas) *Press,* says of the excitement there:

"The news is recent, but it has kicked up a bobbery already which excells the California and Frazer river fevers as much as chain lightning excels a lightning bug. Everyone is crying 'an ox team! a mule team! my kingdom for a pick-axe!!' The plow, the loom, and the anvil, are deserted in many places. Half negotiated trades are broken abruptly off, and every adventurous, imaginative man sees gems in the running brooks, pearls in stones and gold in everything."

---

[58] *Omaha Times,* September 16, 1858. The *Missouri Democrat* of September 15 and the *Boston Evening Transcript* of September 16 carry generally similar stories. The *Lawrence Republican* of September 16 says, "the gold fever continues to rage," tells of the excitement in other cities, and lists "articles constituting a complete outfit for a company of four persons."

GOLD [59]

From the many statements made by gentlemen of veracity and integrity, who are daily arriving in Leavenworth, Wyandott, Kansas City and Westport, direct from Pike's Peak, we are led to believe that there must be gold in that vicinity in considerable quantities. We would not advise our friends to start for the mines before spring – there is certainly more to be lost than gained by such a course. . .

The company which started for Pike's Peak last Monday, brought with them a wagon built expressly for that purpose, by Mr. Weber, of this place. It was, indeed, an excellent piece of workmanship. It is a fact, perhaps not known, that we can turn out better and cheaper wagons than any other town on the river. Parties outfitting for the gold regions can do so in Wyandott at much less cost than elsewhere.

PARTIES OFF FOR THE MINES [60]

We called yesterday morning at the camp of Capt. W. H. Branham, to see him off for Pike's Peak – to say "good by and good luck" to all hands.

They broke camp about nine o'clock, and left in elegant spirits.

Another company, a party of six Germans were camped near by; we called upon them, and said, "wye gastz landsmann" [*wie geht's landsmann* (how goes it, countryman?)], the extent of our knowledge of the German idiom. They had a good outfit, and were feeling as happy as lords, their only regret being, that in the mountains they would be cut short of their allowance of lager beer.

---

[59] *Western Weekly Argus* (Wyandotte), September 16, 1858.

[60] Kansas City *Journal of Commerce,* September 16, 1858. This was the German party of which Count Murat was a member.

More about the Kansas Gold Mines [61]

Corres. of the *Missouri Democrat.*

Kansas City, Mo., Sept. 16.

The excitement about the Kansas gold mines, notwithstanding the many prophesies to the contrary, is still raging. Several companies have already left this point for the diggings, and others are preparing to leave in a few days. One house in this city has outfitted about 30 persons, and others have nearly, if not quite, equaled them. Our merchants look forward to a rich harvest in the spring from those emigrating to Pike's Peak, and appearances justify them in laying in a large lot of supplies for the expedition. I noticed on yesterday an old familiar companion in the shape of a "rocker," for gold washing, sitting in front of one of our tin shops, and in the hopper was placed a tin pan, displaying on the bottom of it the letters, "gold pan," painted in black. This brings to memory some scenes buried with my California life, and almost makes me "sigh for the mines once more." . . Asbury

Off for the Mines [62]

About a dozen of our most enterprising citizens start for the mines, on Monday morning, and although we regret that loss of their pleasant and interesting society, yet we can but admire their energy, and the determined manner in which they set out to give battle to the tediousness of the times.

The season is somewhat advanced for such an under-

---

61 Published in the *Missouri Democrat*, September 21, 1858. The *Omaha Times* of September 16, tells of the departure of a Mr. Kountz and three others for the mines. A letter written from Atchison to the Xenia, Ohio, *Torch Light*, is reprinted in *Freedom's Champion* of October 16, 1858. It tells of the gold fever and says that Mr. Nichols (a member of the Lawrence party) has written from Cherry creek that gold exists there in "the greatest profusion."

62 *Crescent City Oracle*, September 17, 1858.

taking of such a trip, but the stout resolute hearts of the little party will be a match for all the hardships and dangers they may have to encounter, and the heartfelt prayers of many, will quietly arise from the heart for their prosperity, and safe return. The news we have through both public and private sources, leaves no doubt with us, but that we have within five hundred miles of where we now write, gold mines, equal to those of California. . .

PIKE'S PEAK MEETING [63]

Pursuant to call in Monday's *Times,* a meeting was held, corner Delaware and Second streets, to consider the position of Leavenworth in relation to the newly-discovered gold regions. Mayor Denman presided, and, on taking the chair, made a short speech, in which he set forth the advantages of Leavenworth as a starting point for gold-seekers. After his speech, a company of five was appointed to draw up a paper in which the advantageous locality of our city as an outfitting and starting point for emigrants should be duly set forth; said committee to report at an adjourned meeting on Wednesday evening next.

Glorification speeches were made by Judge Purkins, Gen. Larimer, Col. Slough, and a host of other citizens.

The meeting was kept up until a late hour and was the very soul of enthusiasm. We may be sure that Pike's Peak is a living reality, and the main emigration to it must come through Leavenworth.

[63] *Leavenworth Times,* September 18, 1858. Another story in the same paper says parties are starting every day for the gold mines. The *Herald of Freedom* of the same date copies an item from the *Wyandott Gazette* under the heading, "Kansas Gold no Humbug." A dispatch of the 8th from St. Louis, published in the *Boston Evening Transcript,* September 20, says that W. H. Russell will start a train of forty wagons in a few days. *Freedom's Champion* of Atchison, Kansas, in its issue of September 18, tells of the gold fever and gives the advantages of Atchison as a point of departure.

THE SOUTH PLATTE GOLD [64]

Mr. Rupe saw a man by the name of James Saunders,[65] an old mountaineer and Indian trader, just from the gold regions, who reported that there was "plenty of gold." He was at one of the stations above Fort Laramie, and was preparing to go back. He reported that there were a great many in the mines – more than he (Saunders) wanted to see. Mr. G. P. Beauvais, who has a trading post at one of the stations, informed the conductor that he had seen a miner the day before who confirmed the reports of gold. Mr. Beauvais was of opinion that there are rich mines. This is the substance of all Mr. Rupe learned of the gold regions. We shall not probably have anything satisfactory from them until we hear from some of the numerous persons of this city and neighborhood, who are preparing to visit the mines. We have nothing as yet which can be regarded as sufficiently reliable and encouraging to induce emigration.

THE GOLD EXCITEMENT [66]

There is a great excitement throughout the territory about the lately discovered gold in the western extremity of Kansas. Cherry creek and Pike's Peak diggings are about six hundred miles from Topeka. A party consisting of Messrs. Stone, J. Edwards, Morgan Christian, and four or five others from this vicinity left this place last week for the diggings. Another party will start in a few days. We have no reliable informa-

---

[64] This item was published in the *St. Joseph Gazette* of September 18 and copied in the *Daily Missouri Republican* of September 22. The news came from a Mr. Rupe, who had just arrived with the Salt Lake mail.

[65] Saunders was a mountain trader and squaw man. He made trips during the winter between the mines and Fort Laramie, carrying mail and express. See L. R. Hafen, *The Overland Mail*, 145-146. The name frequently appears as "Sanders."

[66] *Kansas Tribune* (Topeka), September 23, 1858.

tion that the mines are rich enough to "pay." Large parties are preparing to leave Leavenworth, Lawrence, and other places.

[OFFICERS FOR ARAPAHOE COUNTY] [67]

In view of the rapid rush of emigration to the gold regions of western Kansas, Gov. Denver [68] has organized Arapahoe county by the appointment of the following officers: Mr. Smith, of Lecompton, probate judge; [68a] J. H. St. Mathew, district attorney; E. W. Wynkoop, sheriff; Hickory Rogers, clerk of board of supervisors. The most of these officers, we believe, are actually en route for the gold region, in a Lecompton company which has just started.

[67] *Lawrence Republican,* September 23, 1858. A. D. Richardson, in his correspondence of September 20 from eastern Kansas, published in the *Boston Daily Journal* of October 4, 1858, tells of parties leaving for the mines and says that a movement is already on foot to create a new territory out of western Kansas. He says that a man from Kansas City on his way to the mines expected to return as delegate from the new territory. *Freedom's Champion* of September 25, 1858, tells also of Governor Denver's appointment of officers for Arapahoe county.

[68] General James W. Denver was governor of Kansas. His appointees had some difficulty in establishing a recognized county government at the mines. These men were instrumental in having the principal town founded in the gold country named for Governor Denver.

[68a] In the *Rocky Mountain News* of December 7, 1870, we find the following data on Judge Smith, picturesque Colorado pioneer: "An item in the obituary column of this paper announces the death of Judge H. P. A. Smith, whom all old settlers will recollect. Judge Smith was a resident of Denver in 1859 and at various times for several years after, but he lived most of the time in Gilpin county. He was prominent in the law practice, in politics, at all kinds of public meetings, and was an amateur actor, Iago being his favorite character. He went to Montana in the early history of that territory. When the vigilance committee undertook to rid that country of 'road' agents the judge rendered himself obnoxious by an attempt to defend some of those gentry, and had to leave Montana. He went to Arizona but returned in a couple of years, when the excitement had subsided, and resided in and about Helena up to the time of his death. He was a man of much more than ordinary ability, but ruined himself and doubtless shortened his life by dissipation and irregular habits."

THE GOLD FEVER [69]

Almost always in every community, there is some object, which, more than all others, engrosses public attention – some times it is scandal, sometimes a tragedy, or something else interesting only in particular localities. At other times it is something exciting more general interest. Here, a few weeks ago, the subject most conversed about, was the "election;" after that was over, it was the "land sales," and the anticipated suffering resulting therefrom; since the land sales have been postponed it has been

Gold! Gold! Pike's Peak! Gold! Gold!

It would be incredulity amounting to absurdity, longer to doubt the existence of gold on the eastern slopes of the mountains, both in this territory and Kansas. Letters have been received from there by citizens in almost every town west of the Mississippi river, all confirming the report of the discoveries. The excitement in every town on the Missouri river, from Sioux City to St. Louis is up to fever heat. . .

[PARTY FROM PACIFIC CITY] [70]

The gold fever has visited Pacific City, Iowa, and our neighbors of the *Herald* say:

A company left Pacific City for the diggings on Tuesday, consisting of H. J. Graham, D. C. Oaks,[71] C. M. Miles, Geo. Pancoast and A. Walroad. This is only an advance guard. . .

A Glenwood, Iowa, correspondent of the same paper says:

The gold fever is raging here, on account of the

---

69 *Nebraska Advertiser,* September 23, 1858.

70 *Omaha Times,* September 23, 1858.

71 See the preceding volume (IX) of this *Southwest Historical Series,* 126-127, for data on Oakes and this party.

recent discoveries of gold deposits at Pike's Peak, near Cherry creek, in Nebraska. One party leaves today with six months' provisions, for that point, and from what I can learn, several more will soon follow. From reliable reports, gold has been found there quite abundant, and persons, with axes and tin pans alone, have been able to realize on an average from five to eight dollars per day, and lumps have this early been found weighing from twelve to twenty dollars.

OFF FOR THE MINES [72]

Our place has been for ten days in a state of excitement, caused by the news from the west, and preparation for emigrating to the new El Dorado.

Yesterday C. H. Blake and A. J. Williams,[73] Esq's., two of our most energetic and prominent citizens, with four well laden wagons and 14 yoke of cattle and two ponies, started for "South Platte," with merchandise to trade with the diggers and mountaineers.

From this place, also went in same company, Messrs. McGlashea, Avery, Willoughby, Gordon, Clark, Conant, and others, making about a dozen who expect to meet companies from Florence, Council Bluffs and

[72] *Crescent City* (Iowa) *Oracle,* September 24, reprinted in the *Omaha Times* of October 7, 1858.

[73] Blake and Williams arrived at the mouth of Cherry creek in late October, and were among the first merchants to set up business in the new country. Blake street, the principal business street of pioneer Denver, was named for Mr. Blake.

Mr. and Mrs. Charles H. Blake lived most of their lives in Denver and Pueblo, Colorado. Mrs. Blake died December 29, 1926.—*The Trail,* xix, no. 9, p. 21.

Andrew J. Williams, a descendant of the famous Roger Williams, was born in Franklin county, New York, November 22, 1833. In 1851 he moved to Iowa, and from there went to Colorado. In the spring of 1859 he and Blake built the "Denver House," first hotel in Denver. Williams subsequently engaged in freighting, farming, mining, and in the cattle business. He died in Denver, May 30, 1895.—Frank Hall, *History of Colorado,* iv, 615.

Omaha at the Elkhorn, making an advance guard to the new mines already near 100 strong. The outfits in teams and materials are of the very best, and complete. The party numbering among its members some of the most energetic and reliable men in this region, many leaving families behind. From Council Bluffs a company of some fifty are already off, and many more will soon follow. Col. Henry Allen, Mr. Graves and Mr. Curtis,[74] three intelligent, stirring men, are among the number, and Captain Smith and Col. Hoops from Florence.

A LARGE SHIPMENT FOR THE MINES [75]

Mr. Samuel Machett, of the firm of Machett, Lindsey & Co., of this city, has left with a large train of merchandise, etc., for the vicinity of Pike's Peak. He designs to locate for the winter at Lupton's old fort,[76] about fifteen miles from the mouth of Cherry creek. Mr. Machett is an old trader, and has an intimate knowledge of the country. He expects to reach his destination in about twenty-eight days. He takes out his usual stock of Indian goods, and will carry on his annual traffic with the tribes of the upper Arkansas, and South Platte. He has promised to send us full particulars in regard to the gold discoveries, and as he has every faculty for communication, we look for intelligence from him as early as the first of December.

HO! FOR PIKE'S PEAK [77]

It is useless to attempt recording the names of even

---

[74] Letters from these men will be given in the next section of this volume.
[75] Kansas City *Journal of Commerce*, September 24, 1858.
[76] This was near present Fort Lupton, Colorado. See L. R. Hafen, "Old Fort Lupton and its Founder," in *Colorado Magazine*, VI, 220-227.
[77] *Missouri Democrat*, September 24, 1857.

river men who have left, are leaving, and contemplate starting for that brilliant center of attraction, Pike's Peak.

We know that Bill Parkinson has been added to the list, and also our humorous young friend, Tom Duncan; and we understand that Oscar Totten,[78] a well known river man, intends going. Others, of a less sanguine disposition, are eagerly awaiting further tidings. . .

IMPORTANT FROM PIKE'S PEAK [79]

The *Wyandotte* (Kansas) *Gazette,* of the 18th says: "Yesterday ten thousand dollars in gold dust arrived from Pike's Peak. One man brought in $6000 as the result of a few week's work. A small boy had $1000, which he says "he dug down and found;" and the little fellow says "he can get all he wants." [80]

These statements are reliable. . .

MORE ABOUT THE GOLD MINES [81]

So great is the demand for provisions at St. Joseph for the mines, that Mr. Isidore Poulin, a merchant of that place, who has been shipping to this port large quantities of bacon, has come here to purchase some of that article. He says that he assisted a few days ago, in carrying from the steamer Wattossa, to the White Cloud, thirty-five thousand dollars worth of gold dust, which Mr. John Richard had procured from the Indians who had collected it with implements of the rudest description, which they made themselves.[82]

Yulo, Yancton, Winnebago, and St. Stephen villages,

---

[78] Letters from Totten will be given in the following section of this volume.
[79] *Missouri Democrat,* September 24, 1858.
[80] This is an example of the unfounded stories that went the rounds.
[81] *Missouri Democrat,* September 25, 1858.
[82] Here is another wild and untrue story.

are points from which many Indians, half-breeds, etc., have gone to the diggings. Mr. Poulin has sold goods for the mines, to the amount of $13,000, and corroborates the reports in regard to the richness of the auriferous region. He is stopping at King's hotel.

FROM PIKE'S PEAK [83]

We have had, of late, contradictory reports from the gold region. A United States officer, just in, reports unfavorably. He says the miners are realizing but little. He attributes this, in a great measure, to the lack of proper tools, but still does not speak confidently of gold deposits. Mr. Zimmerman corroborates the lieutenant's statement in the main, though he thinks there is sufficient gold to remunerate those who are prepared for digging.

Mr. Tappan and Mr. Morrow were in the city yesterday, as agents of the Lawrence company about to start. They bring favorable reports from the Lawrence company now in the gold region. These reports were brought in by Mr. Caldwell, who has been at Pike's Peak during the summer, and returned about a week since for tools and provisions. He reports gold in abun-

---

[83] *Leavenworth Times,* September 25, 1858. This paper, in the same issue, gives a detailed list of an "outfit for Pike's Peak," purchased at Leavenworth by Aaron Jefferies, James Hughes, Aaron Kennedy, and John C. Buell. The total weight was 2900 pounds; the cost $740.32. The *Herald of Freedom* of September 25 says that an outfit for four men for six months will cost about $500, and adds: "We should rather invest the money in one hundred acres of Kansas soil, which will be sure to realize a rich reward."

Freedom's Champion and the *Nebraska City News,* in their issues of September 25, tell of companies forming for trips to Pike's Peak. A letter from Omaha of the same date and similar tone was published in the *Daily Missouri Republican* of October 5, 1858. A. D. Richardson's letter of September 27, from Sumner, Kansas (published in the *Boston Daily Journal,* October 7, 1858), expresses the belief that five thousand persons will be in the mines before spring. The *Nebraska City News* of September 18, telling of a rush to the mines, is quoted in the *Sunday Morning Republican* of September 26, 1858.

dance, and pronounces the region in every respect equal to the representations that have been made. . .

THE KANSAS GOLD MINES [84]

Washington, 27th.

Governor Denver, writing to the Secretary of the Interior, Sept. 17, says that the late news from Pike's Peak leaves no room to doubt the correctness of the reported discoveries of gold in that vicinity. The explorers have found gold on the Arkansas, on the heads of the Kansas, and on the south fork of the Platte river. . .

In view of the present condition of affairs in that region, and to prevent further difficulty, Gov. Denver advises that the lands be withdrawn from pre-emption, if they are open to settlement, leaving them, as in the mining districts of California, free for all who may see fit to engage in working them. . .

It would, he adds, be advisable to send out some competent person to examine the mines, and report the facts in connection with them.

STEAMBOATMEN FOR PIKE'S PEAK [85]

Mr. A. G. B. Baber, who for some time has been traveling as mail agent upon the Missouri river boats, was in our office yesterday and informed us that this was his last trip – that he would return by the South Wester, get off at this city, and start for Pike's Peak, in company with Van Boone, who is to leave some time during the week.

[84] *Boston Evening Transcript,* September 28, 1858.

[85] Kansas City *Journal of Commerce,* September 28, 1858. A correspondent of the *Missouri Republican* writes from Leavenworth on September 28, telling of the stimulating effect the gold rush will have on his city.—*Missouri Republican,* October 5, 1858.

GOLD FEVER INCREASES [86]

Mr. John Huston, one of our oldest citizens, and who for several years past has been more or less on the plains, arrived in this city yesterday evening direct from the gold mines. He says the greatest excitement was manifest all along the route, from Devil's Gap in – that gold has been found on the Medicine Bow, and down the South Platte some eighty miles. Mr. B. Clement (brother-in-law to Mr. H.) will arrive here in a few days, with some $500 in gold, which he took out with nothing but camp tools and a pan; that with a rocker he could have done much better. Some of the men in company with Mr. C. took out from twenty to one hundred dollars a day! Mr. Clement was in the mines only twelve or fourteen days, and succeeded in getting, in that short time, $450 or $500.

Mr. Huston says that at Laramie the greatest excitement prevailed as he came through, on the 5th. Picks and shovels were worth almost their weight in gold. So great was the excitement that the quartermaster at Laramie was compelled to withhold the pay due the hands in the government employ in order to retain them. Most of the mountain traders had repaired to the mines. Mr. H. thinks that in ten or fifteen days large quantities of the gold will be seen here, as many are on their way in. He saw some of the finest specimens of quartz that probably ever has been found in a gold bearing region, and one nugget of gold, weighing

---

[86] Kansas City *Metropolitan*, September 29, and reprinted in the *Missouri Statesmen,* October 8, 1858. This story, with less detail, appeared in the Kansas City *Journal of Commerce* of September 29 and in the *Missouri Republican* of October 4, 1858. The *Council Bluffs Bugle* of September 29 tells of departures for Pike's Peak. The *Journal of Commerce* of October 23, 1858, tells of the arrival of Ben Clemoor (Clement), brother-in-law of Huston.

twenty-three ounces, taken out by a Mr. Jackson,[87] formerly of Glasgow, Mo. There were thought to be 300 persons in the mines on the first of Sept. Most of those who were high up in the mountains had been driven down into the valleys by a snowstorm, which fell nearly three feet on the 5th.

FIRST GOLD FROM PIKE'S PEAK [88]

Gold from Pike's Peak. – We saw yesterday, a lump of gold, weighing over three ounces, resembling in shape and quality the California gold as it comes from the diggins, which was sent to Mr. Cook, of the firm of Cook & Matthews, of this city.[89] The curious may see the specimen at the jewelry store of C. D. Sullivan, opposite the Planter's House.

OUTFITS FOR THE MINES [90]

We call the attention of all persons going to the Kansas gold mines to the very important fact that Topeka is the last place on the route to the "diggings" where a good and complete outfit can be obtained. Teams of all descriptions, wagons, and to the numerous merchandise houses we call particular attention. Flour, tea, coffee, blankets, miner's boots, picks, pans, blowers, hardware, tinware, and all description of dry and West India goods can be had at Topeka, at a price very little in advance of river prices. Flour, meal, corn, potatoes, and all kinds of produce can be had here at a less price than at the Missouri river. The flouring mill at this

[87] This was George A. Jackson, later famous as the discoverer of gold at Idaho Springs in early January 1859. See L. R. Hafen (ed.), "George A. Jackson's Diary, 1858-1859," in *Colorado Magazine*, XII, 201-214.

[88] *Missouri Republican*, September 29, 1858.

[89] Marshall Cook was in the mines in the fall of 1858.

[90] *Kansas Tribune*, September 30, 1858. The *Western Weekly Argus* of Wyandott, Kansas, September 30, says that "merchants are almost overrun with customers outfitting for the mines."

place is doing a heavy business, and is making flour of a superior quality.

Miners, buy your outfits at Topeka, and save the expense of hauling 75 miles from the river. . .

GOLD MEETING [91]

Pursuant to public notice, a very large and enthusiastic meeting was held in Overton's hall, Wyandott, on Saturday evening, September 18th, 1858, "for the purpose of taking measures to inform the world of the nearness of Wyandott to the Kansas gold mines, and to transact any business connected with the interests of the city, in view of its proximity to the new El Dorado."

W. P. Overton was called to the chair, and R. B. Taylor appointed secretary.

Dr. J. P. Root was called upon to state the objects of the meeting, and did so in some pertinent remarks, and went on to give the reasons why the people of this place should take such action as would inform the eastern emigration of the great advantage of making Wyandott the outfitting place for those going back to the new mines.

The chairman urged the necessity of the people taking some energetic measures to bring to the notice of the world the advantages of Wyandott as a commercial point.

Mr. J. M. Winchell followed in a spirited speech; and in the course of his remarks, referred to the route for a railroad up the Kansas river. He said he had been over the route, and thought it the finest and most feasible route for a railroad that he had ever seen, and that it was in the direct course to the gold mines. He thought the travel to the mines would do much toward

<hr>

91 *Western Weekly Argus*, September 30, 1858.

paving the way for the early building of the road. . .
[A long report of this meeting and an adjourned one
follows.]

PIKE'S PEAK GOLD FEVER [92]

It seems from more recent and more reliable accounts than those we published last week, in reference
to the late magnificent discoveries of gold in the vicinity
of Pike's Peak, Kansas territory, that the whole affair
is an unmitigated humbug. The company that went
there from this county last spring have returned; and
to them we are indebted for this piece of information.
That there is gold in the vicinity of Pike's Peak, cannot
be questioned. And that it has never been found there
in sufficient quantities to remunerate one for the time
and labor expended in its procurement, is equally certain. The base impostors who fabricate such groundless
stories as have been going the rounds of the press, for
the last few weeks, in relation to the gold discoveries
referred to above, merit the severest censure. Thousands
of the credulous have been deceived by their false
statements, and induced to sacrifice positions of comfort and profit to join an adventure – hazardous at best
– in the vain hope [of] attaining immediate independence. Let the credulous beware.

FROM PIKE'S PEAK AND NEW MEXICO [93]

It will be remembered that E. R. Zimmerman returned to this city from the gold region while the fever
was at its height. He had been prospecting the whole
summer and told a straight-forward story. But as his
representations did not coincide with the general feel-

---

[92] *Richmond Mirror,* October 2, 1858. This item was supplied by Floyd C.
Shoemaker, secretary of the Missouri State Historical Society.

[93] *Leavenworth Times,* October 2, 1858. This story was reprinted in the
Daily Missouri Republican, October 4, 1858.

ing they were received with distrust and suspicion. Later evidence substantiates everything asserted by Mr. Zimmerman. Mr. Spaulding arrived in the city from the gold district on Sunday and pronounced "the diggins" considerable of myth. Though gold was everywhere found it could not be obtained in any quantity sufficient to justify the labor necessary to obtain it. Mr. S— made a thorough survey and only left when he became convinced that it would be more than useless to remain.

James Miller arrived in our city yesterday from Pike's Peak, direct. He left the Cherokee country last spring with a company of fifty-five, and not only prospected the entire gold district but crossed over into New Mexico. His company gave the various diggings a thorough trial and became convinced that no paying deposits could be found. He crossed over to Sante Christo [Sangre de Cristo mountains] and spent a few weeks in New Mexico. Had a pleasant chat with the celebrated Kit Carson who did not credit the gold reports, and finally returned home with the Missouri company with about one dollar's worth of the precious ore.

Mr. Miller reports that most of the miners determined on leaving the country and that they all felt much disheartened.

We give these reports as we have them that our readers may hear all sides. We shall never allow ourselves to be the means of creating a false impression, or of misleading the public.

ANOTHER VISIT FROM CANTRELL [94]

An article appeared in the Cass County Democrat, last week, embodying the statement of B. G. Johnson,

94 Kansas City *Journal of Commerce,* October 2, 1858.

just returned from Pike's Peak, in which ungenerous reflections were made in regard to Mr. Cantrell and his report condemned as intentionally untrue. Mr. Cantrell called upon us yesterday and desired us to state that he sought no newspaper controversy, and did not know that his statements were to be made matters of newspaper notoriety. But in order to leave no uncertainty he reiterated his statements, which [we] give below:

He left Fort Laramie on the 23rd of July, having heard that gold was being found on the South Platte, and being but a few days out of his route he concluded to go and see. It thus appears from Johnson's own statement that he left Pike's Peak on the 25th, which must have made a distance of three hundred miles between them at their nearest proximity. So much for Johnson's statement that Cantrell never saw the mines.

On the last day of July, Cantrell crossed Cherry creek, and remained in that vicinity until the fifth of August, at which time he started home, with his wagons and teams.

He found above the mouth of Cherry creek, from 4 to 15 miles, a portion of the Georgia company, under Green Russell, who had prospected that distance and were still at work. Russell, himself had gone up to the mountains on a prospecting tour, and was not expected back until some days after Cantrell left.

FOR THE GOLD FIELDS [95]

A company of fifty persons have left Council Bluffs for the Cherry creek gold fields. Other companies are outfitting.

---

[95] *Nebraska City News,* October 2, 1858. In the same issue of this paper the *Nebraskian* is quoted in telling of emigration to the mines from Omaha. Correspondence of October 2 from Fort Kearny reports parties en route to the gold fields.

[Another item in the same paper:]

On Tuesday, Messrs. Jerome Dauchy, John K. Gilman and Isaac Sager started as an express to the gold fields. They took two horses and a mule each. They expect to be absent but about sixty days.

THE NEWEST GOLD REGION [96]

A correspondent asks us to advise him and other mechanics of this city, who have a little money and a very slim prospect for work, to start for the new gold mines in western Kansas. We answer:

Gold-digging is the very last resort to which we would impel industrious young men, especially those who have good trades. Better dig gold than stand idle, but better plant corn, shoe horses, build houses or make fences than do either of these. There are brilliant prizes in the gold-digging lottery, but there are many blanks to each of them; and the young man who can earn $100 per annum over the cost of his board and clothes should do it rather than court the risk of goldhunting.

But if our correspondent and his friends have an average of $50 each and no assurance of work here through the winter, we do most earnestly advise them to set their faces toward Pike's Peak at once, but without intending to see that elevation before next May. Take passage at once to Chicago, Davenport, Keokuk, Burlington (Iowa), or St. Louis; and there inquire for work; if none is to be had, push on to Jefferson City, St. Joseph, Leavenworth, Atchison, or Lawrence, and inquire further, determined to find work somewhere this side of Smoky Hill fork for the winter. If you cannot obtain money for your labor, take board and lodging, store-orders, anything that you can live on, and thus worry through the winter with as little loss and

---
[96] An editorial in the *New York Tribune*, October 2, 1858.

as much gain as possible. When spring opens you will very probably find work on better terms on the Mississippi or the Missouri; if not, you will be within striking distance of the new gold field, and can set your face toward it as soon as grass will serve. Very likely, however, so many will be going that you can do better by staying. That there is much gold this side of the Rocky mountains is scarcely questionable; but you are quite likely to acquire a share of it more easily in almost any other way than by digging it.

As to proceeding to Pike's Peak this season, it is not worth a moment's consideration. You would meet bitter blasts from snowy mountains ere you could reach it, and find neither food nor shelter at your journey's end. Too many have started already.

LETTER FROM KANSAS [97]

Sumner, K.T., Oct. 5, 1858.

From the gold mines the intelligence has been conflicting since I wrote you last. Some parties have come in who have been quite successful, and bring flattering reports, while two or three others who have arrived express the conviction that the "diggings" will never pay, and that the whole affair is a grand humbug. Companies continue to leave almost daily for the mines. Everything indicates that they will meet with much suffering on the journey. Latest reports from the west inform us that heavy snows have already fallen there.

OFF FOR PIKE'S PEAK [98]

A company of about a dozen of our citizens started

---

[97] Extract from the regular correspondence of A. D. Richardson to the *Boston Daily Journal* and published in its issue of October 16, 1858. The *Missouri Republican*, in its issue of October 5, gives a discussion of conflicting reports coming from the mining region.

[98] *Nebraska Advertiser*, October 7, 1858.

yesterday for the gold regions. They were supplied with implements for digging gold, and with sufficient provisions to last about eight months.

On Tuesday last a train of five waggons and fifteen men, from Rockport, Mo., passed through this place for Cherry creek.

ROUTES TO THE GOLD MINES – OUTFITTING POINTS, ETC.[99]

We notice that a good deal of spirited discussion is going on between the Leavenworth and Kansas City papers, about the best routes to the mines. The Kansas City folks, as is natural, contend that the great Santa Fe road via Council Grove, and the valley of the Arkansas, is by far the best route; while the Leavenworth papers contend no less earnestly that the route up the Kansas river via Fort Riley and the Smoky Hill fork, is the best one. Undoubtedly each route has its own peculiar advantage, and probably, on the whole, there is no very great difference in them. The Kansas valley route, however, keeps much farther in a settled country than does the Santa Fe route. There are settlements on the Kansas river and on the Smoky Hill, for a distance of two hundred to two hundred and fifty miles. Phillips is located near the junction of the Smoky Hill and Saline forks, and there are sparse settlements for fifty miles above him. Of course, such a route would be much more favorable for obtaining feed for cattle and mules, than one through a country entirely destitute of settlements, as is the Santa Fe road almost the entire distance from Council Grove.

We notice, too, a great rivalry among all the river

[99] *Lawrence Republican,* October 7, 1858. In the same issue this paper quotes from a number of its exchanges, giving their reports of departures for the gold fields. Several of these are reproduced above.

towns, each claiming to be the best outfitting point for companies going to the mines. We take a little different view of the matter. It would seem to us that the best point for outfitting should be, that point nearest the mines where a good outfit can be obtained, and the best choice of routes be had. . . Apply these principles, and Lawrence possesses advantages for outfitting which no other town in the territory can claim. In the first place, most articles for an outfit can be obtained here as cheap, or even cheaper, than at Leavenworth or Kansas City. . .

Looking at the matter, then, dispassionately, our advice to parties from the east and elsewhere would be, as Lawrence is the extreme western point where an outfit can be obtained, and presents a choice of routes so that the best one can always be taken, to make this the headquarters for starting. From the Missouri river, Lawrence is always accessible by daily lines of stages.

ANOTHER WHEEL-BORROW MAN [100]

In 1849 or '50, some sprig of young America left this city for California, transporting all his outfit in a wheelborrow. He got through safe, and made trip in much quicker time than most of the ox trains.

On Wednesday last, Mr. Alex O. McGrew, of Pittsburg, Penn., left this city for Pike's Peak – solitary and alone – taking his provisions, blankets, pick and shovel, gun and ammunition in a wheelborrow; and we have not the slightest fear that he will be successful, and make the trip in good time. When he left Pittsburg, he had just five cents in money and no outfit save the raiment he had upon his back. A company from that city

[100] Kansas City *Journal of Commerce,* October 8, 1858. We shall hear more of McGrew, "the wheelbarrow man."

had left a few days before him, and it was his intention to overtake them, and go in their company.

He, however, arrived here ahead of the company, (his party has not arrived yet,) and being impatient to get off, made arrangements to travel with a company, just ready to leave from Wyandott. On the eve of their departure, however, the sheriff called upon them to audit some few financial vouchers he had in his possession, which threw McGrew upon his own resources again.

Nothing daunted, he goes to Mr. George E. Pitkin, hardware dealer in this city, with whom he has been acquainted for several years, and from Mr. Pitkin gets his wheelborrow and outfit, bids his friends "good bye," and starts on his pilgrimage of over six hundred miles through an uninhabited country. Between this city and Westport, he was overtaken by our fellow citizen, J. C. McCoy, Esq., who was then outfitting a party for the Peak, and admiring his zeal and ambition offered him a place in his company, free of charge.

Mr. McGrew thanked him very heartily for his proffered kindness, adding that he could not consistently accept the offer, as he was in great haste, and feared that he might be delayed, and perhaps caught in the snow, if he traveled with oxen! . . .

PARTIES OF GOLD HUNTERS [101]

The *Spread Eagle* came into port yesterday morning with some two hundred passengers and a large cargo of freight. Among her passengers were three different parties for Pike's Peak.

One company of four persons were to go out with a party of the American Fur Company's men, who will

[101] *Ibid.*, October 8, 1858.

leave their camps some time next week. Another company of four, got off at Wayne City, and will start with some of their friends from Independence. Another party of seven got off at this city, viz. Josiah Smith, Stephen Smith,[102] John Elliott, John Coons, William Swan, and two young men from Leavenworth City. This company came well prepared with a traveling outfit, and will remain in the city for four days, making purchases for their provision, camp and mining outfits. This company are prepared to go out in good shape, and make express time, having a fine lot of horses and a most superior ambulance, for service on the plains, made by Messrs. T. B. Edgar & Co., of St. Louis. Most of the company are young men, who have been engaged in business in St. Louis.

FIRST LETTER FROM THE GOLD HUNTERS [103]

Council Grove, K.T., Sept. 24.

Messrs. Van Horn and Abeel: Dear Sirs: According to promise I write you from this point respecting our journey thus far, and the prospects awaiting us.

We left camp four miles this side of Westport, on the morning of Thursday, September 16th. Our train was then composed of six wagons with twenty-five men and one lady, who was attending her husband to the new Eldorado.

At Indian creek, where we "nooned," we were joined by the two Wyandott companies, composed of two wagons with six men each. We camped the first night near Olathe, a beautiful and enterprising village of about three hundred inhabitants. Here we organized,

---

102 These were among the founders of Fountain City (site of present Pueblo, Colorado) in the fall of 1858.

103 Published in the Kansas City *Journal of Commerce*, October 5, 1858.

electing Mr. John I. Price captain, and Messrs. Winchester and Gross, of Kansas City, assistants, whose duties were well defined and specified in an article of agreement which had been framed before leaving Kansas City.

From thence we proceeded to McCamish, some fifteen miles distant, where we were again joined by companies consisting of twelve men and two ladies, from Lafayette county, near Lexington.

At Rock creek, some eight miles beyond One Hundred and Ten creek, we came up with Dr. Woodward, and the Rev. Mr. Linderman, formerly of Natchez, Miss. They had one wagon and four men.

At One Hundred and Ten creek we were overtaken by two wagons from Kansas City, with eight men, headed by Messrs. Rogers and Kellogg,[104] and at this point we found awaiting our arrival, companies from Topeka, Tecumseh and Lawrence, so that now we have a train of some nineteen wagons, with perhaps 100 men and three ladies – every wagon well supplied and provisioned for the journey.

Thus far we have progressed finely, losing none of our stock, and all the boys being in fine spirits. We have been met by several wagons returning from Pike's Peak, whose reports are rather discouraging. They showed us some specimens which they brought from the mines, of the very finest quality, but in very small particles, . . .

104 Kellogg's diary of this trip was published in *The Trail*, December 1912 and January 1913. *The Trail* of May 1917 runs an obituary of David Kellogg. From it we obtain the following information: He left home as a youth and engaged in steamboating on the Mississippi. He became an ardent abolitionist and fought with John Brown in Kansas. From Colorado, Kellogg went to the Pacific coast. He settled in Seattle in 1863, and thereafter made that city his home.

Our boys, however, are nothing daunted or dispirited, all are fully determined to push on and prospect for themselves before returning. Nothing else will satisfy them. . . Very respectfully, HAMPTON L. BOON.[105]

ANOTHER LETTER FROM THE GOLD HUNTERS [106]

Oct. 6, 1858

Messrs. Van Horn & Abeel: Dear Sirs: I wrote you on the 24th ult., from Council Grove, and write again from this point to give still further items of our trip. . .

At Pawnee fork we met two gentlemen who were returning to the states, who both informed us that they conversed at Fort Union with a Mexican who had just come in from the Peak and reported that he had found gold in abundance, showing them the finest specimens. He brought with him some $700 in cash, and intended taking back supplies for emigrants this winter.

Another gentleman informs us that he found gold at every point on Cherry creek where he prospected but in small quantities, hardly sufficient to pay the miner, although he admits that he had no mining implements.

Another report says that the miners refuse to work when they fail to get less than 25 cents to the pan. Our old California miners, Messrs. Price and Parkinson, tell us that it is equivalent to $100 per day.

To counterbalance these reports, I must say that we have met with those whose statements are rather discouraging, all admitting, however, that there is gold there. . . Respectfully yours, Hampton L. Boon

---

[105] Hampton Lynch Boone, a great-grandson of George Boone (brother of Daniel Boone), was born in Fayette, Missouri, December 15, 1837. He attended Bethany College at Wheeling, West Virginia, studied law at Lebanon, Tennessee, and practiced law for a time. During the Civil war he was a lieutenant in the Confederate army under General Price. He died at Ardmore, Indian territory, April 8, 1893.—H. A. Speaker, *The Boone Family*, 284.

[106] Kansas City *Journal of Commerce*, October 19, 1858.

CROSSING THE PLAINS

From a contemporary sketch of 1859

[LETTERS OF G. N. WOODWARD EN ROUTE
TO THE MINES][107]

Council Grove, K.T., Sept. 23.

Dear Sir: We arrived here last evening about sunset, in good health and spirits. Near us is a band of about 400 Kaw Indians, and this morning our camp is filled with squaws, papooses and Indians. They are a miserable lowlived set, who live by thieving. We have to keep a guard to prevent their stealing our cattle.

We heard good news from the mines yesterday. Men there are making from $10 to $15 per day.

We go on slowly, but hope to get through in thirty days from Kansas City. There are sixty men and twelve wagons in the company, all well provided with provisions, ammunition, etc.

I will finish this at Diamond spring, twenty miles west of here.

Elm creek, Sept. 23.

Camped for the night. We are now fairly out of the plains. There is no habitation in sight or sign of civilization – but prairie – boundless, endless. I feel first rate – free, free as air! We live by the side of our wagon and sleep in the tent. I do the washing, Charlie washes the dishes, and Dunton drives the team and attends to the oxen and wagon. I write this in my tent, on my knee, I cannot get at my writing paper, and am obliged to write this on leaves torn from my memorandum book, so you must excuse the brevity of this letter.

Diamond spring, Sept. 25.

There is no postoffice here so I shall send this by a gentleman who is going to Westport.

To-day we leave the settlements for good. Two teams joined us here, one from Lawrence and the other from

---

107 Kansas City *Journal of Commerce*, October 8, 1858.

Topeka. We are all well as yet, and in good spirits. This will be the last you will hear from me till I get to the Arkansas crossing or Bent's Fort. Yours truly, G. N. WOODWARD.

Big Coon creek, October 1, 1858 [108]

Dear Sir: We are now two hundred and forty miles from Kansas City, in the midst of buffaloes, wolves, Indians and game of all kinds. We have lain in camp today to find two yoke of oxen which strayed from us. We found them just at night.

Yesterday we saw buffalo for the first time. I shot the first one shot by the company. To-day they have killed five. The meat is very coarse and tough. I shot some ducks this morning which were excellent.

We saw a party yesterday from Fort Union. They report the news from the mines good. . .

While I am in my tent writing by the light of my lantern, the Germans are singing, and the others are fiddling and dancing; so you see we have merry times out here.

Oct. 7.

We are now thirty miles from Arkansas crossing. I have no time to write, except to say that we are all well. Yours truly, G. N. WOODWARD.

SUPPLY TRAINS FOR THE CHERRY CREEK MINES [109]

We understand that the enterprising and indefatigable firm of Russel, Majors & Waddell are fitting out two supply trains of twenty-six wagons each, for the Cherry creek mines.[110] Fifty-two wagons loaded with

---

108 *Ibid.*, October 19, 1858.

109 *Nebraska City News,* October 9, 1858. The same paper copies a story from the St. Louis *Evening News,* telling of a lump of gold sent to Mr. Cook from his brother in the mines.

110 Apparently the trains never made the trip.

supplies will help support the miners in that region, and prove a fair remuneration to the enterprising gentlemen who conduct the affair. Nebraska City must inevitably be the grand starting point for the new gold fields; and we are happy to see the great movement so happily inaugurated as in the enterprise just noticed. Of course emigration can but take this natural and best route – the route of the national firm of Russell, Majors & Waddell. Let the emigration come. It will here find good outfits at reasonable prices.

[Other items from the same paper:]

A company from Brownville has been in our place during the week fitting out for the gold mines.

Fifty-two passed through our city on Wednesday for the gold fields.

There have been several companies in from Missouri outfitting at this place for the gold fields. Come on – to the best route, and to as good outfitting post as any other on the river.

PIKE'S PEAK [111]

Four companies have been fitted out this past week, for the gold regions. Wilson and Company furnished the hardware – Meyers & Applegate their stoves and tinware. The companies are: Trace & Wood, Fisher & Co., B. F. Miller & Co., Staggs & Co., Larimer & Co. There are two or three more companies in town fitting out – two of them are from Oscaloosa, Jefferson county.

GONE TO THE GOLD DIGGINGS [112]

Mr. Ed. Cook, Bicknell, Buckwalter and Hotchkiss started for the gold "diggins" from our city last

---

111 *Leavenworth Times,* October 9, 1858. Another article in the same paper tells of Captain Humphrey and his company en route for the mines.

112 *Kansas Weekly Press* (Elwood), October 9, 1858.

Wednesday. They were provided with tools, provisions, etc., and have gone with the determination to give the new El Dorado a fair trial. Mr. Cook promised to send us for publication, a correct statement about the mines, on his arrival at Cherry creek.

Mr. Mitchell, Corley, Cheatly and others also started for the "Land of Promise" on Thursday, and we are informed that another company is fitting out and will start in a day or two. We wish them all a pleasant trip and "whole acres of gold."

We are only 660 miles from Pike's Peak and are consequently the nearest point in eastern Kansas. Elwood is the only city in Kansas from which the Salt Lake mails depart and arrive weekly, and the only place from which parties leaving for the gold region are always sure of the company and protection of large government trains.

***

LARGEST OUTFIT TAKEN OUT [113]

By the *Sioux City* yesterday morning there arrived in Kansas City, a company from Washington, Mo., bound for Pike's Peak. The company is under the charge of Messrs. Ming & Cooper,[114] and consists of ten persons, with six wagons, thirty yoke of oxen and 25,000 pounds of freight. These men are old travelers on the plains, and go out fully prepared to make every edge cut in the new mines. They take the great Santa Fe road to the Arkansas, thence to Bent's Fort, Puebla, and the mines. They go well prepared for the coming winter, and will no doubt give a good account of themselves. We wish them every success.

---

[113] Kansas City *Journal of Commerce*, October 13, 1858.

[114] Ming and Cooper became merchants in pioneer Denver. We shall hear more of them. The issue of the preceding day told of the return of David Kendall from Fort Union and of his meeting gold seekers. His advice regarding relations with the Indians is given.

More Gold News – Letter [115]

We clip the annexed letter from the *Cincinnati Times* of October. It is the latest news received, and still further confirms the existence of gold in the newly discovered mines. Our latest dates were to Sept. 3d, brought by John Huston. This is eight days later. The *Times* vouches for the credibility of the writer:

Direct from the Gold Regions.

The following letter has been handed us for publication. It comes direct from the new gold region, and is addressed to a friend in this city. Its statements can be relied upon.

Pike's Peak, Sept. 11, 1858

Friend Dennis: I left Salt Lake on the 22d of August. When I got to the South Platte, eight men and myself came out here to the gold country, and I am glad we came. I have made thirty dollars and over since I got here. You had better come out here as soon as you can, by way of Fort Riley. Come on this fall; I think the chance is best. You know what will suit the plains as well as I can tell you. If you can get out flour and bacon to last till spring, it will be better for you, because provisions are high. I cannot write any more; the man is going this moment that takes this letter. From your friend, EDMOND D. TYNE.

What have the reprint editors to say to this letter? Is it written by some one connected with traders and Missouri river merchants? We have a thorough contempt for this whole class of scribblers, who have neither the industry to read, nor the capacity to comprehend the facts by which they are surrounded. . .

Another Departure for the Gold Mines [116]

We announced a few days since, the departure for

115 Kansas City *Journal of Commerce*, October 13, 1858.
116 *Missouri Democrat* (St. Louis), October 13, 1858.

Pike's Peak of five young men whose names we gave, and stated that others would soon leave this city for the purpose of joining them. The Grey Eagle company of three, vis: Arthur McCoy, tinner, Wm. Bates, blacksmith, and Thomas Farmer, painter, left this city on the *E. A. Ogden,* and will unite with the others at Kansas City.

BEST ROUTE TO CHERRY CREEK [117]

It is quite amusing to see the innumerable places mentioned as the nearest and best point from which to "roll out" for the newly discovered "gold fields." Now we believe something like romance is being practiced upon those who intend going out to the mines.

We would like to know how Omaha and Kansas City can each be the points for emigrants to start to Cherry creek. Omaha, if we are not mistaken, is about seventy miles north of Brownville, and Kansas City a long way south of this. Yet both of these towns claim to be nearest to the new Eldorado. – Will ye enlighten us, gents?

We do not claim superiority over any other point on the river as a starting point for the gold placers in the Pike's Peak region. All we ask is to examine your maps; look at your locality and the locality of the mines, and the truth will be apparent to a man "without eyes." – Don't have the mud rubbed in too thick. Look out!

GOLD BULLETIN [118]
(From our special gold correspondent.
Council Grove, Sept. 27, 1858)

EDS. REPUBLICAN : – Well, we have arrived at Council Grove. We left Lawrence Sept. 23d; so you will see by the date of this that we have been four days be-

---

[117] *Nebraska Advertiser* (Brownville), October 14, 1858.
[118] *Lawrence Republican,* October 14, 1858.

tween the two points. There has nothing occurred so far on our route that would be of interest to your readers. We have had beautiful weather, passed through a fine country, and still are in camp on the borders of civilization, and are in hopes that all will continue as pleasant until we arrive at the grand terminus – Cherry creek.

The Santa Fe mail arrived last evening, bringing the news up to Sept. 10th, fully corroborating all the accounts of the gold diggings. A gentleman who came through informed me that the whole of New Mexico was in one gold stampede; that men who were doing a good business at home, were leaving everything and going to the new mines.

The Wyandot company are some three days ahead of us; so the probabilities are we shall have to depend upon ourselves for protection. To-day we have been in camp cleaning up our old muskets, some of which were the trophies of the late wars.

Sheriff Wynkoop, Judge Smith and Hickory Rogers are all on their way, armed with commissions from Gov. Denver for the purpose of organizing Arapahoe county, and informing the people that President Buchanan has had communication with Queen Victoria. All this will be edifying to the Indians in that vicinity.

Well, supper is ready, and I must close. It presents rather a ludicrous scene to see men that heretofore have been surrounded by all the luxuries of life, at the sound of the cook's voice rush to the spot where the tin dishes are scattered over the ground, denoting the place where every man is to draw his rations. . . WM. O'DON-NALL.[119]

119 Mr. O'Donnall wrote another letter en route, from "Fort Atchison" on the Arkansas river, October 7. It was published in the *Lawrence Republican* of November 11, 1858. He mentions Captain Rogers and Sheriff Wynkoop as members of the party, but gives little information of interest or value.

THE PLAINS AND THE MOUNTAINS – INFORMATION [120]

Since the first reports of the discovery of gold reached us, we have often had occasion to furnish various details and statements respecting these discoveries.

In almost every instance, our information has come direct from men who have been for years trading, freighting, or traveling over the plains and in the mountains. Indeed, it is very seldom that a day goes by that we are not visited by some of these mountain and prairie men, either for the purpose of furnishing us with additional gold information, or to give in the particulars of their last trip, or to give us extracts from their letters, received from their business agents, either in the mountains, at the forts, or upon the plains. Among these men we can mention the names of the following gentlemen, who are well known as freighting contractors over the plains: Kitchen, Bent, Campbell, Hatcher, O'Neil, Yager & Kerr, Hunter & Scott, Harris & Scruggs, and Hays, Cantrell & Rickman.

We may also refer to the following, who have been for a long time employed in the capacity of wagon masters: Wilson, Kendall, Fleming, Vogel, King, Hickman, Perry, Long, Ferrill, and Brodwell.

We say that nearly every day, we are visited by some of these mountain men, who always come in with a kind of practical information for the benefit of men who wish to ascertain geographical or business particulars, that never have and never will be furnished by the agents or corps of the various bureaus and departments of the government.

---

[120] Kansas City *Journal of Commerce,* October 16, 1858. The *Kansas Weekly Herald* of October 16 tells of the arrival of a Mr. Williams, "an old mountaineer," from the mines and says that he will return in about ten days.

GOLD FROM PIKE'S PEAK [121]

Mr. A. C. Hopkins, of this city, has left in our possession a specimen of peculiar dark colored sand stone, in which very minute particles of gold thickly abound. The stone is so soft it may be crushed easily by compression with the fingers, making the work of extracting the auriferous matter a cheap process. Mr. H. Collins, who owns the specimen, and who brought it from Pike's Peak, will leave it at this office for a few days.

OFF FOR PIKE'S PEAK [122]

Mr. Bates, Mr. Palmer and Mr. McCoy, were passengers yesterday by the E. A. Ogden, having left St. Louis for Pike's Peak. These young men are all mechanics, and go out to the mines prepared to make money at the forge and anvil, with square and compass, as well as at digging. Mr. McCoy has an intimate knowledge of the mines, having "been there before," as he informed us. They leave to-day by the Santa Fe route.

GOLD! GOLD! [123]

ARRIVAL OF MEMBERS OF THE LAWRENCE COMPANY DIRECT FROM THE MINES: THE FORMER ACCOUNTS CONFIRMED: PLENTY OF GOLD ON THE SOUTH PLATTE:

We are at last enabled to lay before our readers,

[121] *Daily Missouri Republican,* October 16, 1858. The *Nebraska City News* of the same date quotes the *Leavenworth Herald* in telling of the arrival of gold seekers from Newark, Ohio: "They say the greatest excitement prevails in that portion of Ohio, and most every one had the gold fever." A. D. Richardson's correspondence of October 20 (published in the *Boston Daily Journal,* November 2, 1858) tells of the latest news from the mines via the Salt Lake mail. The report is that one hundred men are at the mines and that they are averaging $10 per day.

[122] Kansas City *Journal of Commerce,* October 19, 1858.

[123] *Lawrence Republican,* October 21, 1858.

direct and reliable news from the newly discovered
gold fields of western Kansas. Mr. F. H. Brittan arrived in this city last Sunday evening [October 17],
direct from the mines, having been just twenty-eight
days on the route. Mr. Brittan has lived in the territory
for nearly two years, residing most of the time at Burlington, Coffey co., at which place he and his father
were for some time proprietors of a hotel. Mr. Brittan
left Burlington last spring, about three weeks after the
Lawrence company left this city. He joined the Lawrence company at Pike's Peak upon his arrival there,
and has been with them ever since. Mr. Brittan confirms the accounts we have from time to time published
from the company. The most of the summer was spent
by them in prospecting over a large extent of country,
finding gold in almost all cases, but not, as was thought,
in paying quantities, except by one prospecting party,
who found on the head-waters of the Platte diggings
where they thought from five to eight dollars per day
might be made. [Then follows an account of the experiences of the Lawrence party, as described in the
preceding volume of this series.]

Mr. Brittan intends to remain during the winter,
and will return to the diggings next spring with a full
outfit for mining. The remainder of the Lawrence
company intended to remain in the mines about two
weeks after Mr. Brittan left, and then come down the
South Platte about forty miles, to Bent & St. Vrain's
Old Fort,[124] and winter there. The weather was fine
at the mines when he left, but with very cold, frosty
nights; and it was the opinion of old mountaineers, that
within a few weeks the miners would all be obliged to

---

[124] St. Vrain's Fort, on the east bank of the South Platte, about a mile
north of the mouth of St. Vrain's creek.

seek winter quarters – some at Fort Laramie, others at Fort Bridger, and some had returned to the states.

There were about 100 persons in the mines, and numbers arriving daily, when Mr. Brittan left. Provisions were tolerably plenty, though at enormous rates – flour selling at $25 per hundred. The gold dust is used in the mines as currency, at one dollar per pennyweight.

Mr. Brittan met from 700 to 1,000 emigrants en route for the mines. . .

The implements in use in the mines, are, almost without exception, only shovels, picks and pans. It was the general opinion of the miners, and among them several old Californians, that, with the proper implements, from five to fifteen dollars per day could be made. Mr. John Rooker [125] had a rough rocker made out of a hollow log, split, with which he and his son were taking out from four to eight dollars per day. A half-breed Cheyenne boy, John Smith by name, son of John S. Smith,[126] well known as an old Indian trader, had been at work about two weeks when Mr. Brittan left, and had made fourteen dollars per week with his pan, shovel and pick, carrying his dirt from ten to twenty rods. One "mess" found in prospecting a "pocket" in the bed rock, where the dirt yielded a dollar and a half to the pan-full. One of the Georgia company washed out sixteen dollars in a single day. . .

For heavily loaded teams, Mr. Brittan thinks the route from Lawrence to Council Grove and the Arkansas would be the best, though about one hundred miles farther than the northern route. The shortest and best route for those going on horseback, he thinks,

---

125 For data on Rooker, see the preceding volume (IX) of the *Southwest Historical Series,* 197.

126 *Ibid.,* 116-117.

would be from Lawrence up the Kansas river, via Fort
Riley and the Smoky Hill fork – saving probably a
hundred miles over either of the other routes.

Capt. Smith and Wm. B. Parsons, Esq., arrived in
town on Monday morning, by the Topeka stage. . .[127]

[KETTLE OF GOLD] [128]

After the recent conflicting reports, we take pleasure
in laying before the readers of the *Press* undoubted
evidence of the existence of gold in large quantities
on our western border. On Friday, Mr. A. M. Smith, –
a gentleman known to Mr. Brace, our postmaster, and
to other parties here and in St. Joseph, as a man whose
testimony can be relied upon – arrived in Elwood from
Nemaha county. Mr. Smith has lately seen a kettle
of gold brought by his friend Mr. Robinson from
Cherry creek valued at from $6,000 to $7,000. Mr.
Robinson only left Pottowattomie county in May for
the gold regions. He was thirty days in going there,
and has returned in 28 days – although he had to make
a road for himself some portion of the way. He went
well prepared for working in the mines and had two
men to assist him. The three were about two months
in obtaining the dust which he has brought back. . .

[127] An account follows of a meeting held in Lawrence Monday night. It
was addressed by Smith and Parsons, who gave encouraging reports of the
gold fields. Captain G. W. Smith (son of Judge Smith of Lawrence, who was
elected governor under the Lecompton constitution) wrote a letter on October
30 (published in the *Border Star* and copied in the *Kansas Weekly Herald*
of November 20, 1858). He gives an account of the experiences of the Lawrence party and a favorable report as to mining prospects. Parsons prepared
accounts of the experiences of the Lawrence party. These were reproduced
as Appendices F and G, in the preceding volume of this series. Parsons's
guidebook was also reprinted in that volume.

[128] *Kansas Weekly Press,* October 23, 1858. This "kettle of gold" story
appears to be a complete fabrication. The *Lawrence Republican* of November
4 copied this story from the *Weekly Press.* The *Crescent City Oracle* of
October 22 tells of a party bound for the mines.

We have become so accustomed to the sight of the emigrant wagon and the "prairie schooner," that we hardly turn to look at or question the emigrants who arrive here every day. They come from Iowa, Wisconsin, Illinois and other states, and come so constantly that those states already witness the losses they have suffered. The ferry over the Missouri at Elwood hardly makes a home trip without one or more of these emigrant wagons, with the droves of cattle and mounted drivers, which always come with them. Leavenworth and Lawrence, get more of the "carpet-sack" emigrants, but Elwood, situated at the western terminus of the old California road, and the present Utah mail line, and opposite St. Joseph, Mo., is the great thoroughfare of the overland Kansas emigration.

GOLD BULLETIN [129]

The Elwood *Weekly Press,* of Saturday the 23rd ult., has the following additional cheering intelligence from the new Eldorado: [then follows the "kettle of gold" story printed above.]

For Pike's Peak

Notwithstanding the lateness of the season, trains after trains continue to throng the road, no longer, however, bound for Laramie or Salt Lake, but for the new El Dorado. Not a day passes that quite a large number of wagons do not arrive and depart from this

[129] *Lawrence Republican,* November 4, 1858. The *Nebraska Advertiser* of the same date runs an editorial giving the advantages of Brownville (opposite St. Joseph, Mo.) as an outfitting point and urging the necessity of building a road from that point to the mines. The *Omaha Times* of the same date quotes the *Leavenworth Journal* in telling of the return of George W. Vincent from Utah, by way of the mines. He is reported to have obtained $57 in three days of digging. The *Boston Transcript* of November 4 quotes the *Herald of Freedom* as saying that "not less than 1000 persons will winter in the vicinity of Pike's Peak the coming winter," and that a movement is on foot to divide Kansas and organize a new territory.

point; in fact, emigrants to the gold diggings have become so common that it is useless to ask them where they are bound, as all invariably return the same answer. – *Palmetto Kansas,* (Marysville.)

News from the Gold Hunters

We have been permitted to copy the following extract from a letter received by Mr. Badolett, of the Union hotel in this city, from the Steamboat gold hunters, Oscar Totten, Capt. Parkinson, Duncan, Noble and their companions. Their friends will be glad to hear that they are getting along finely.

Big Bend of the Arkansas, October 12, 1858

Dear Sir: – We have had a capital time since we left Kansas City, but have met few incidents worthy of note. One or two stampedes of our mules, and the good luck to kill three buffalo since we came into the buffalo range, is about all we have had for excitement, or to vary the monotony of a trip across the prairie. We have had no additional news from the mountains since we left Westport. Your friends Totten, Duncan, and the others send their respects. All well. Yours, etc. CHAS. H. NOBLE. – Kansas City *Jour. of Com.*

FROM THE PLAINS – GOLD NEWS. – From the *Independence Dispatch* of the 27th ult., we learn of the arrival in that city of Mr. O. Allen,[130] who has been acting as general guide for the Utah army; from which we make the following extracts:

An exploring party from Arkansas was met the 4th of September on the Big Medicine Bow. This party had gone up the Arkansas river taking in route for Pike's Peak, Cherry creek and South Platte, thence up Cast Lapodo [Cache la Poudre], thence through

---

[130] Mr. Allen was the author of one of the guidebooks, description and summary of which were published in the preceding volume of this series.

the Laramie plains to the point where Mr. Allen met them.[131]

They had discovered gold as far as the tributaries of the Medicine Bow. They represent the country in the vicinity of Pike's Peak and Cherry creek, as similar to Grass valley in California, and the neighborhood of Nevada City. The specimens of gold which Mr. Allen saw were of a very rich quality, and he expresses his belief that the mines will turn out well.

BOUND FOR THE PEAK. – Let the reports be what they may, the people from all parts of the country are either preparing to leave, or leaving for the gold mines of Pike's Peak. Through our town of Junction City have passed several parties. On Saturday, the 16th inst., three wagons, heavily ladened, with ponies, etc., containing all the necessary equipments of a mining party, passed. The names of the gentlemen comprising the party are, W. H. Green, Daniel Davis, Joseph M. Green, William H. Wignall, Ahija Chambers, Emery Williams, J. A. Bruce, George Cowle, O. H. Parshall, Wm. J. Adair, John Kearns and Samuel Harrison. On Monday, the 18th inst., another party went through, consisting of the following named individuals: Capt. D. D. Cook, James Hall, W. G. Reid, J. H. Russell, H. Humphrey and J. McCormick. These gentlemen were from Kansas, Missouri, Kentucky, Illinois, Wisconsin, Michigan, New York, etc.

Although reports from this gold country are not so favorable as at first, still the adventurous spirit of our young countrymen induces them to explore for them-

---

131 This was the Green Russell party, on its tour into present Wyoming, as related in the preceding volume (*Southwest Historical Series*, IX, 114-116). This item identifies the military party whom the prospectors met and with whom the miners shared their bear meat.

selves, and either become satisfied of the humbug from personal examination, or convinced by getting a pocket full of rocks of the supposed richness of those parts. If nothing more is received than has yet come to hand, we for one will be tempted to try our fortune in that part of the territory. – *Junction City Sentinel.*

FOR PIKE'S PEAK [132]

A company of "Pike's Peakers" passed through Lexington, Mo., a few days ago. They were Dr. G. R. Melton, Nathaniel Patten, G. W. McElhaney, Robert McElhaney, and Drake Marlow, of Montgomery county; James Yosti, of St. Charles county; and J. Donnelly, of Louisville, Ky.

Dr. Melton has consented to become a correspondent of this paper from Pike's Peak.

THE WHEELBORROW MAN [133]

We learn from Mr. David Morris, of the house of Hale & Bro., who has just returned from the territory, that the wheelborrow man who left here a few weeks ago for Pike's Peak, taking his outfit in a wheelborrow, passed several ox trains, and at Council Grove overtook the party he set out to go with. He traveled from twenty to thirty miles per day, and was everywhere well received and entertained by the settlers free of charge. His spirit and energy could not fail to make friends wherever he went. We expect to hear from him as among the first who have made their "pile" in the new gold mines.

[132] *Missouri Statesmen,* November 5, 1858. The same paper tells of Mr. Robinson and his kettle of gold.

[133] Kansas City *Journal of Commerce,* November 6, 1858.

EN ROUTE FOR PIKE'S PEAK [LARIMER'S LETTER] [134]
Little Arkansas, K.T. Oct. 14, 1858

Editor of the Times: As your readers may feel an interest in the welfare of their citizens on the way to Pike's Peak, and having leisure this afternoon, I dot you down a few items.

Our party, now consisting of eight wagons and thirty-two men, are getting along first rate; still we have taken a round-a-bout way. Your people should at the opening of spring, urge the opening-up of the Fort Riley and Republican fork route, it is so very direct from Leavenworth City to the gold diggings. The Santa Fe road is the best natural road in the world, and to compete with it you must open-up your own natural route up the Republican fork.

The boys are all in good spirits, and enjoy camp-life very much. So far, we have had no delay, and are now inside of three hundred miles from Bent's Fort. Getting across the ferry at Topeka was the only place that gave us trouble; still we were delayed only a few hours. One day we traveled thirty miles, but from fifteen to twenty miles per day is getting along well.

We laid up one day at Topeka waiting for Mr. Wade, but still we had to leave him, as he had not then found his oxen.

We find no trouble in keeping up with mule teams

---

[134] *Leavenworth Times,* November 6, 1858. General William Larimer was to become a principal founder of Denver. One of the streets carries his name. He was of Scotch descent and was born in Westmoreland county, Pennsylvania, October 24, 1809. He became an active abolitionist and was prominent in Pennsylvania politics. In 1855 he moved to Nebraska, where he shortly became a member of the legislature. He later recruited the Third regiment of Colorado volunteers for the Civil war and became its colonel. After the war he returned to Leavenworth, where he died on May 16, 1875. *Reminiscences of General William Larimer,* etc. (Lancaster, Pennsylvania, 1918) contains several of his letters and much data on the gold rush of 1858-1859.

with our oxen, and my opinion is, after thirteen days' trial, that oxen are every way better suited for the trip than mules. . .

We are now in the buffalo country. Yesterday we had a splendid hunt, and captured a fine one. You have no idea of the excitement of a buffalo chase. . .

We have heard but little from the gold diggings. A train from Fort Union reports favorably. We expect to reach Bent's Fort in about fifteen days from this. . . WM. LARIMER.

FRESH ARRIVAL FROM THE SOUTH PLATTE GOLD REGIONS! [135]

During the week there have been several arrivals in Lawrence direct from the Cherry creek gold diggings, bringing dates to the 1st of October. The following members of the Lawrence company have just come in: Messrs. M. A. French, Wm. Smith, J. D. Miller and A. Voorhees.[136] They each bring with them rich specimens of gold, and fully corroborate the news brought in by Messrs. Parsons, Smith and Brittan. They left the mines on the 1st of October, some ten days after the gentlemen above named. As it may be interesting to know what further explorations were made subsequent to the leaving of Parsons and the others, we will give a statement of facts as obtained from Mr. French.

The company were still prospecting, and had continued to find gold in quantities varying from 10 cents

---

[135] *Lawrence Republican,* November 11, 1858.

A. D. Richardson wrote from Lawrence, Kansas, November 11 (letter published in the *Boston Daily Journal,* November 20), telling of the return of members of the Lawrence party. He had interviewed William Hartley, Jr., formerly of New Haven, and his report was similar to that printed here, from the *Lawrence Republican.*

[136] These were members of the Lawrence party. For their experiences consult the preceding volume in this series.

GENERAL WILLIAM LARIMER

to $1.50 per panful. The latter amount was only found occasionally. Some of the company had gone up on to Dry creek, a stream emptying into the Platte ten miles above the mouth of Cherry creek, and had met with good success, the average yield per panful being 37½ cts. Mr. Rooker, who was mentioned in Mr. Brittan's statement as working with a rude rocker made of a hollow log split, was still taking out about $4 per day. He is quite an old man, somewhat feeble and works slowly. Mr. Hartley of Wabonsa took ten pansful of dirt out of the place where Rooker is digging, and washed from it $4.75, by weight. The old miners there, those who have been in California and Australia, think that the "show" is a good one. The Georgia company, who, it was supposed, had gone to Fort Laramie for the winter, came in a few days after Parsons left, and reported that they had been on a prospecting tour along the base of the mountains for some hundred miles north, finding gold in about the same quantities as at Cherry creek. They are so well satisfied that they intend to remain in the mines during the winter, and were about despatching a team for provisions to New Mexico when Mr. French left.

The weather was still mild and pleasant in the mines. The stock will be driven down the Platte some thirty miles, to be wintered on the bottoms, while the mass of miners will remain in the mines during the winter. There were about fifty men in the mines when Mr. French left. Up to that time none of the eastern emigration had arrived, and there was no knowledge or expectation of it at the diggings. The first immigrants were met about 100 miles this side of Cherry creek, and consisted of seven men and two horse teams, and were from Iowa. The next day a company of three men, with a single wagon drawn by mules, was met. They were from Leavenworth, but Mr. French did not learn their names. No more were met until this side of the Platte crossing of the Laramie road, from which point to Fort Kearny trains were constantly met, consisting of companies from Council Bluffs, Omaha, Iowa Point, Elwood, St. Joseph, etc. The trains were generally in good condition, but had been delayed some by stampedes of their stock. The returning company experienced no bad weather, . . .

Mr. French and his companions intend to return to the mines next spring.

The health of the men continued good. Mr. Churchill, who had been wounded in the hand by the accidental discharge of his gun, had recovered.

WEDNESDAY, Nov. 10. – Messrs. Hartley,[137] Bradt, Churchill, and Dickson, arrived in Lawrence, last evening, from the Cherry creek mines. They were in company with Messrs. French and others, most of the way in, and fully corroborate his statements in regard to the mines. They are very enthusiastic, and will return early in the spring.

REPORT OF MR. SIMPSON [138]

Mr. Geo. S. Simpson, an old mountaineer, a man who has lived under the shadow of the Rocky mountains for over eighteen years, called on us yesterday morning, and gave us some additional information about the gold regions consisting in part of advice to emigrants, description of route, etc. etc.

---

[137] These men were of the original Russell and Lawrence parties. A "Statement of Wm. Hartley, American Miner," dated at Lawrence, November 18, and published in the *Lawrence Republican* of November 25, 1858, gives a story similar to that of French, Parsons, and other members of the returning party.

[138] Kansas City *Journal of Commerce*, November 13, 1858. See volume IX of this series, page 199, for a sketch of Simpson.

Mr. S. says that he has known of the existence of gold in the mountains about the head waters of the Platte, Arkansas, and Colorado or Green river, for more than fifteen years. He says, that so far as he has been able to learn, the miners, though they are now quite successful, have not been prospecting in the right quarter to find large deposits of gold. Mr. S. being of the opinion that gold exists more plentifully in the mountains stretching southward from the Arkansas, than in the range north of this stream.

THE GOLD EXCITEMENT [139]

The most prominent name, amongst the returned gold hunters of Cherry creek, is that of our old friend W. B. Smedley,[140] Esq., of Ray county, Missouri. He it was who got up the company and explored those diggings from last June until the middle of September. It was he who wrote the first letter we ever saw from those parts, exciting the hopes and desires of the people, but it now appears that his first letter and conclusions were incorrect. Below we publish a letter directly from him, and add that every word from him can be safely relied upon. Our belief in the existence of gold in sufficient quantities to pay well the worker, is not at all shaken by these late reports; for with the unfavorable comes most flattering accounts from those parts. We submit them all to the reader, and expect him to use his own judgement in forming his conclusion. – *Junction, Kan., Sentinel.*

RICHMOND, RAY CO., Mo., Oct. 10th, 1858.

FRIEND KEYSER: – Mr. Elder has just written to you, and I take this opportunity to add a few words to what

---

139 *Herald of Freedom* (Topeka), November 13, 1858.

140 See Appendix E, in the preceding volume, for letters from Smedley with an account of the gold seekers from Ray and Bates counties.

he has said. I have just returned from the alleged El Dorado of K.T., not as you might infer from current reports, with my "pocket full of rocks," but far different, having obtained but one dollar. Now, I do not intend to say that this is all, or even an average of what a person might be able to do in a season. Neither will I say, positively, that one cannot make five or even eight dollars per day, but I wish to give you a simple statement of facts. Early last spring a company from this place started for Pike's Peak, and upon the road we fell in company with fifteen others from Bates county, Mo. When we arrived at South Platte we met with Capt. Beck, C. [G.] Hicks, of Indian territory, and a party from Georgia, making in all 103 men, all intent upon finding the shining ore. After dividing our forces into small parties of some six or eight, we went into the mountains; and after an absence of some ten days we returned into camp, and the report of each company was the same, (no gold, not even the color.) In the meantime, those who remained in camp, prospected the streams in the immediate vicinity. Long toms and rockers were used, but still we were not able to make anything. We then removed to the famous Cherry creek again, (having previously traversed the entire distance of this creek, and finding a little better prospect than in any other place) for the purpose of further prospecting, which we did to our satisfaction in a few days, when we started en route for home, having made up our minds that what gold we could obtain would not be worth working for. We came to Pike's Peak, where we met the Lawrence boys, who persuaded us to remain a short time longer, which we did: and while with them we again made an effort to prospect the mountains, but with no better success, when we started home again, parting from the Lawrence boys at that place. But after traveling 50 miles south, we were overtaken by a team from Taos, and from reports which they gave us were induced to go to Ft. Massachusetts,[141] where we found the best prospects upon the whole route, still not enough to induce us to remain. We came home, overtaking some of the Lawrence boys, who also were coming home, and from them we obtained a history of what they had done in our absence, which amounts to this: After remaining at Pike's Peak for fifteen days, and before going to Cherry creek (from which some of their letters purport to be dated) they started home, those who were coming, and others pushed on to California, their destination. When arriving at Huerjana [Huerfano], and finding that we had gone over the mountains, some followed and others came home. Now, sir, this statement you can rely upon as correct. Whether those, then, who tell us of their being out there this summer, have told us the truth or not, I do not say; but where could they have kept themselves during our stay there? Where are the evidences of their work? Could not the number of men who were out there this summer, and were scattered throughout the country, have met them or seen traces of their having been there? Or have they been there since August the 10th? Part of our company remained there until that time. Some tell us that their friends have told them they made $600 and upwards in one week. Who are their friends? Let us hear from them, or are they ashamed to show themselves to the public? Some may be of the C. C. Carpenter stripe, who is now going through the country peddling brass specimens, some of which I have now in my possession; he has 60 oz. of the same. Query, is the $800 lump in

141 Fort Massachusetts, founded in 1852, was the predecessor of Fort Garland (Colorado) and was located six miles north of the latter.

Leavenworth off from the same piece? It strikes me that one who can manufacture such enormous tales to order, could also substantiate their statements by manufactured specimens? What are their motives? One man may have an interest in 1500 yoke of cattle, which must be sold. Another wishes to sell a large lot of provisions and get well paid for freighting the same. It would be well for the public to be a little cautious about the bait, however well it may be prepared.

Now it is not only wrong, but absolutely wicked for them to deceive the people so; they are daily sending men, women and children into a very inhospitable climate, and with but few provisions. What their conditions will be it is hard to contemplate. I do not purpose either to encourage or discourage any one who wishes to go next spring; let them understand that they have diggings yet to find. There is a little fine gold there, there may be more, I know not, but my opinion is that it will turn out nothing but a humbug; and in this opinion, most who have visited the mines will coincide. Very respectfully, yours, WM. B. SMEDLEY.

WORTHY OF CONSIDERATION [142]

We publish in another place, this week, a letter from Mr. Smedley, who was the principal in getting up the Missouri and Cherokee company last spring to Pike's Peak, written by Mr. S. to the editor of the Junction *Sentinel,* a newspaper located at the town of that name, at the junction of the Republican and Smoky Hill fork, in Kansas. His statement is very unfavorable. We publish it, that all sides may be heard, and that our readers may have the means of arriving at the truth in the premises. We are all in want of gold, and we sincerely hope it may prove abundant at Pike's Peak; and yet

[142] *Herald of Freedom,* November 13, 1858.

we must not shut our eyes to the fact that conflicting statements are made by those who have been there. Truthful men represent that fair wages can be made in the mines; and men whose characters for veracity are equally good, say that gold does not exist there in sufficient quantities to pay for working.

We are conscious that if we would fill up our columns with favorable news from that region, and follow it up from week to week, we could sell any number of copies of our paper, and could make a good thing of it. Thousands who read our columns, and know we mean to be truthful, would rush to Kansas, and we should be enriched by the rise of town property; but there is another thing to be considered. By our caution, and refusing to exaggerate in the past, we have established a character for our journal as a truthful one, and that character we are not willing to hazard by running off in any of those wild adventures. True, we might do as others have done. As soon as there are indications of the excitement subsiding on any question which engrosses the public mind, we could labor to get up another subject; and another, and another, from year to year, so long as the public would be "bamboozled" by us; but we were taught in youth, that "Honesty is the best policy," and are resolved to be guided by it, let it land us where it will.

No man has read the conflicting accounts from Pike's Peak without doubts crossing his mind whether gold will ever be found in sufficient quantities to make it an object to go there to delve for it. That gold is found there, we think there is no doubt. So gold was found last summer in Iowa, and pearls were found on the Walnut, in Kansas; but the former would not pay for mining, and the latter were nearly valueless.

Messrs Smith and Parsons have given the public the

most impartial statement which has been laid before them in regard to the gold region, but when their account is analyzed, doubts, instead of being removed, are confirmed.

Nearly fifty persons went out in a body last spring to explore those regions, determined to find gold if it existed there. They have spent a whole season in the search, and just as winter is setting in, three of them returned. Their stock of gold – their summer's wages – their hopes of the future are all preserved, not in a heavy bag filled with gold, but in a goose quill, and the quill is not full. We may not be correctly advised in this regard, and will be happy to make corrections if we are in error, but if we are correct, this speaks volumes upon the subject.

The future is expected to develop a more favorable condition of things; but who is there who can afford to pay $25 a hundred for flour, and for other necessaries corresponding prices, spend one month in going to the mines, one month in returning, and be to such an enormous cost in getting there, and when there, not able to labor over four months in the mines during the season, and then make but "four or five dollars per day?" The whole winter must be lost from the inclemency of the climate, owing to its great altitude, and all the discomforts in a superlative degree must be borne through the whole period the adventurer is absent from the settlements: and all for "four or five dollars a day!"

A majority of those who have gone to the Peaks, and who are preparing to go, would feel indignant to be offered a dollar a day and board for a year's services; and yet, of those who have returned, or those who will return next fall with their constitutions broken with hardships, how many will have cleared above expenses twenty-five cents per day?

Though we do not intend to throw cold water upon those who are adventurers, and wish to explore that region, and whose hopes of golden realities are ardent, yet we do wish to present to them for consideration, a correct view of the matter, so that, should they be disappointed, the disappointment will not fall so severely upon them as it would under other circumstances. . .

THE TRUTH IN REGARD TO THE GOLD DIGGINGS – PARSONS VS. BROWN [143]
"WORTHY OF CONSIDERATION."

Correspondence of the Lawrence Republican.

MR. EDITOR, – The very sapient editor of the *Herald of Freedom* wasted a column of his last issue by filling it with an article headed, with more than doubtful propriety, as above. The article appears like a "goodly apple, rotten at the core;" for while the outside reflects to the public, who are uninitiated in the mysteries of that worthy's inner temple, an ardent desire to serve the people, and a philanthropic determination to rescue them from the horrors of humbuggery, the hidden meaning, or rather motive, stands out as plainly revealed to those who are acquainted with the author's peccadilloes as Envy's hateful form ever does.

He states that he publishes Mr. Smedley's letter in regard to the gold mines, in order that his readers "may have the means of arriving at the truth in the premises" – italicising the word "truth" with his characteristic sarcasm, which plainly means that the accounts heretofore published by myself and companions are not reliable means of arriving at the truth – in other words that we lie. Now I suppose that Brown [144] might tell me that I lie – nineteen times, more or less, and I

---

[143] *Lawrence Republican*, November 18, 1858.
[144] Editor of the *Herald of Freedom*.

should not get very mad about it, but be very much amused at his foolishness; and I do not write this article to show my resentment, but rather to exhibit his stupidity.

He says: "Men whose characters for veracity are equally good, say that gold does not exist there in sufficient quantities to pay for working." I say it does exist in sufficient quantities, to pay for working. I say that it does exist in sufficient quantities, etc., for I have seen it and tested it; and Mr. Brown, with, perhaps, an equally good character for veracity, says "it does not exist," etc.; and let an intelligent public decide between us. The public will say that Mr. Brown, who has never been west of the settlements, knows nothing about it; and that I do. Multiply Mr. Brown by a thousand, and in any court my evidence would still outweigh them. Patrick was arraigned for an assault — the offence was clearly proven by two men who saw him strike the blow, and he was taken to the lock-up muttering curses loud and deep against the "joodissery" because the court wouldn't hear the testimony of thirty or forty men who didn't see him strike the blow.

My position is simply this: that not a single man has returned from the mines that has had a fair opportunity of examining them who does not pronounce them rich, and each new arrival brings more favorable reports.

Mr. Smedley left us at Pike's Peak, as he says, on the 25th of July – compelled to leave, as it were, by the discouragement and shiftlessness of most of his men. I know Mr. Smedley well, and I know he was inclined to remain; and I further know that I could explain, to the perfect satisfaction of Mr. Smedley, all of our subsequent movements, which he seems to consider a mystery. Mr. Smedley and company remained in the vicinity of Cherry creek but a few days, and seems to have forgotten that they left behind thirteen Georgians who were satisfied that the mines would pay, and refused to come back with Mr. Smedley, and who made them pay from that time forth, are still there, and making good wages.

Mr. Smedley asks: "Have they been there since Aug. 10th?" Yes, Mr. Smedley; we arrived there Sept. 2 – one month and nine or ten days after you left for the states, during all of which time the Georgians were developing the mines.

The above will serve to show how fair an opportunity Mr. S. and company had to know concerning the mines. He left a month before they had been fairly tested; and no man has returned since they were fairly and energetically worked except members of the Lawrence company and Messrs. Peers and Clenmore, and these all pronounce the mines rich.

Mr. Brown's brain has become muddled by the confused rumors brought by correspondents of California papers, returned camp-followers from Salt Lake and Laramie, none of whom have been within 180 miles of the mines, and, as far as concerns the value of their information, might as well have been returning from Labrador. The people of Lawrence have seen five persons from the mines while the people of Laramie have seen one. Men who were in that section two, three or even one year since, have no more right to say there is no gold there than a captain of a whaler who put into San Francisco bay thirty years ago for supplies would have to say there is no gold in California. The erudite editor remarks further, that he "might fill his columns with favorable news from the mines" – and he undoubtedly might do so if he would take it second-handed, but, in my opinion, he doesn't publish the accounts because he didn't get the accounts to publish; and thus, with a great show of regard for truth, and with his characteristic shrewdness, he makes a virtue of necessity. Brown does a great many sly things, but they are awfully patent.

After indulging in some reflections concerning the reputation of his journal for veracity – which exhibit a great depth of originality and vividness of imagination – and recalling some lessons of his youth for long years forgotten, he refers to the amount of gold which Messrs. Smith, Brittan and myself brought back, and parades it in italics and capitals as a striking proof that the mines amount to nothing – totally ignoring the explanation which he has had of the matter, and which would satisfy a tolerably intelligent mule.

Brown knew, when he penned the article, or ought to have known if he knows anything that is going on around him, that by walking fifty rods he could see $100 in dust brought from the mines; and yet he studiously conceals the fact. He seems bound to sink everything which he cannot pilot through, even though he has to trust himself with the others to the tender mercies of a plank. He cooly asks how men can afford to pay $25 per hundred for flour, and make only $4 or $5 per day. Our outfit cost less than $150 per man, for six months. Four or five dollars per day will amount in that time to $800, thus leaving $650 profit – which I hope is within the astute financier's comprehension.

And then again he talks of "broken down constitutions." This is decidedly rich. He informs our friends east who may by accident pick up his paper, that we are the ghosts of our former selves, when here we are – fatter, stronger and healthier than ever before in our lives; and all those that we left behind are the same.

But I can waste no more time or space on the subject, having already wearied your readers too long; but I do hope that Brown will learn that because he can see no good in a thing, it does not follow that there is positive evil – and be reminded that to doubt everything, is by no means an unerring mark of a wise man. Yours, WILLIAM B. PARSONS.[145]

FROM THE GOLD MINES – RELIABLE INFORMATION [146]

The following letter, though not of very recent date, will be read with interest, because relating to a subject uppermost in the minds of thousands. The information here given is from an intelligent and reliable source, and is worthy the credence of all. We say this to strangers and others abroad; in this city Mr. G. P. Beauvais is known to hundreds. The letter is written to Mr. A. Beauvais of this city.

LONE STAR, N.T.,[147] October 16, 1858

DEAR BROTHER: Herewith I send you a specimen of gold dust from the South Platte mines, by a Mr. Geddings from Ohio, who has just arrived from the mines.

From him I gather the following facts, which are mostly in contradiction of the various publications circulated in the papers. He states that there have been about thirty men at work on the Platte, and thinks that the whole amount of gold taken at the various places which have been worked will not exceed $1500, and also that the various stories about taking it out in large or heavy nuggets are base fabrications, circulated by ignorant persons. He also states that he is acquainted

---

145 See Appendices F and G in the preceding volume for additional letters from Parsons.

146 *St. Joseph Gazette*, November 16, copied in the *Daily Missouri Republican*, November 20, 1858.

147 On the North Platte, near Fort Laramie.

with the principal part of the miners who have prospected the country the past season, and thinks that nine-tenths of all the miners who left the mines for the states, either had to buy or beg specimens to take with them, or else to go without any. He also states that the largest piece that has been taken out anywhere between the Spanish Peaks, and the Medicine Bow mountains, weighed just eleven grains, and was taken to St. Louis by a Mr. Pierce, and that the best mines found as yet are on the South Platte and Cherry creek, about eighty miles northeast of Pike's Peak. With very hard labor men have been making from one to four dollars per day, with rude tools constructed on the ground.

One party of Georgians (Green Russell's party) whom he thinks are the only party of experienced miners who have been prospecting the country, have made the best wages of any who have worked – they making from four to sixteen dollars per day, with the use of proper tools and quicksilver, and he (Mr. Russell) thinks that good wages can be made by those who wish to work hard.

The mountains have not yet been prospected, and the more experienced miners are of the opinion that more extensive deposits will be discovered next season. The gold at present discovered, is extensively scattered over the country, but very fine, and in localities where the almost entire lack of water renders it impossible to work it until the spring rains come on. The country, aside from the gold, is valuable, and desirable for agricultural purposes, as well as for raising and feeding stock.

The above is a correct and true copy of Mr. Y. S. Geddings' statement regarding the mines to me. G. P. BEAUVAIS.

ACCOUNT OF MR. S. S. SMITH [148]

Mr. S. S. Smith, who went from this city on the first day of October last, for the Pike's Peak gold mines, returned on Sunday, being forty-four days making the trip. He left the mines on Cherry creek, on the 30th of October. Mr. Smith is well known to many of our citizens. He has been in the California mines for some years, and has perhaps traveled further the past season in search of gold than any man on the continent. He left California last summer for Frazer river, remained for some time at the Colville mines,[149] but was driven away by the Indians, crossed the country and the Rocky mountains to the head waters of the Missouri, descended the river in a canoe to Fort Randall, where he found the steamer *D. A. January,* on which he shipped to Kansas City. Arriving here about the period of the first gold excitement, he determined to visit Pike's Peak, and started on the first of October, and reached his destination on the 27th; on the 28th he went out prospecting upon the head waters of Cherry creek, working only four hours with a shovel and pan, and obtaining between four and five dollars in gold dust. Obtaining such information from the miners as he deemed of importance, he started back the next day. . .

The miners are agreed that from five to twenty-five dollars per day could be easily made on any of the streams of that section of the country, with the proper

---

[148] Kansas City *Journal of Commerce,* November 16, 1858.

The *Kansas Chief* of November 18 says the gold mines are largely a humbug. The *Lawrence Republican* of the same date says a letter received from its correspondent reports that O'Donnall and companions had reached Bent's Fort on October 22 and at that point met members of the Georgia party direct from the mines with $200 in gold dust. The *Boston Evening Transcript* of November 18, quotes the *Cincinnati Gazette* with a letter from Kansas telling the story of three miners arriving with $6000 or $7000 in gold.

[149] In northeastern Washington.

mining apparatus. As yet, but very little prospecting has been done, and that very little poorly.

Chub creek, Medicine Bow fork, Cherry creek, and a few other streams, comprise the whole prospecting circuit of the miners now there. . .

All through New Mexico, preparations were being made to take large quantities of provisions in the spring – also very many Mexicans were preparing for mining operations. All the miners were in good spirits and intend to remain in the mines – the Indians prophesying a mild winter.[150]

CONFIRMATION OF THE GOLD REPORTS [151]

We had the pleasure, last evening, of a conversation with J. Bradt and S. C. Dickson,[152] who have just ar-

[150] Apparently Smith's report was copied in the *Missouri Democrat* of St. Louis. Regarding the story and its publication the *Missouri Republican* of December 2 said: "The *Democrat* felicitated itself, a few days since, in having been ahead of this paper with the news it published, some days ago, from the Pike's Peak gold region, brought in by Mr. S. S. Smith.

"To all the credit it deserves for hunting up that news it is perfectly welcome. As we are now advised, Mr. S. S. Smith left Kansas City on the 11th of October, and proceeded no further than the crossing of the Arkansas. Having at that point met the Santa Fe mail, he returned with it. The story recounting the rapid journey of this member of the Smith family to Cherry creek—his four hours' experience in mining, the results being $4 worth of dust, or $1 an hour, etc., may be set down as a pleasant little romance, which Mr. Smith was kind enough to devise for the gratification of border inquisitiveness on the *qui vive* for mining news. This story was, however, paraded in the *Democrat* as a sober narration of doings in the diggings. It had a quizzical air, but was nevertheless in that respect a great improvement upon that marvellous news about Walker in El Paso, which treated the readers of the same paper to a *feu de joie* of laughable fibs, and has not been excelled by any performance in the same line, since the respected correspondent, who once haunted the locality of that paper, the 'Intelligent Delaware Indian,' left town."

[151] *Leavenworth Times,* November 20, 1858.

[152] Members of the Russell, or Georgia, party. See accounts in the preceding volume (IX). From Lawrence, W. B. Parsons wrote a communication to the *Boston Evening Transcript* (published in the issue of November 23), telling of the return of the party of fifteen, which left the mines October 1,

rived from the gold region of Kansas. They came in the company of fifteen, which left Cherry creek, Oct. 1st., an account of which we gave in a late issue.

Both of our informants, who are estimable and reliable gentlemen, concur in representing the gold in the vicinity of Pike's Peak as abundant, and scattered over a large area of country. They propose returning early in the spring with a large company.

Our informants have been prospecting during the summer, and Mr. Dickson informed us that he would not have worked at ten dollars a day.

Two towns have been started in the gold region – Montana and St. Charles.[153] Montana is located on the Platte river, about eight miles from Cherry creek, toward the mountains. St. Charles is at the junction of the Cherry creek and Platte rivers. A large number of settlers are farming, and extensive improvements will soon be made. The health of the gold region was unequalled, and all the miners were in good spirits. Settlements were met fifty miles west of Fort Riley.

LETTER FROM OSCAR B. TOTTEN [154]

In camp, Forty miles from Bent's Fort, Oct. 26, 1858

Mr. Editor: I embrace a favored opportunity of sending you a few lines while on my way to Pike's Peak.

---

about ten days after Parsons left. This latest intelligence, Parsons points out, corroborates his previous reports. The Kansas City *Journal of Commerce* of November 23 tells of the arrival of French, Smith, Dickson, and Hartley and gives their favorable report of the mines. The *Daily Missouri Republican* of November 27 announces arrival of the party at St. Louis the day previous. It gives a brief account of experiences of the Lawrence party and a favorable picture of developments at the mines. J. A. Churchill, a member of the Lawrence party, wrote a communication to the *Boston Daily Journal* (published in the issue of December 22, 1858) telling of his experiences at the mines.

153 Parts of present Denver.

154 *Missouri Republican*, November 24, 1858.

We are getting along well and our company are in good spirits.

We this morning had the pleasure of meeting three gentlemen – two Mr. Russells and one other [155] – just from the mining region, who report very favorably. They are returning to get supplies, and will start back in the spring. They say the country, as yet, has not been prospected thoroughly, and as far as their knowledge extends, it will without doubt prove a very rich mining country. . .

The following are the names of gentlemen belonging to our company: Wm. H. Parkinson, Oscar B. Totten, Charles H. Noble, Albert Baber, Wm. Wimer, John Winn (?), Daniel Rabuny, Henry McCarty, V. McCarty, Isaac Harrington, Jacob Harrington, Green Allen, Henry Dorsey, Richard Rahuny, Richard Chafley, and many others, which I have not time to write. When we arrive I will post you in regard to the diggings. Yours, etc., OSCAR B. TOTTEN

[ROUTES TO THE MINES] [156]

It is interesting to look over the papers published at different towns on the Missouri, below Omaha. Every town that can boast of three houses, a well and a smoke house, are showing up their advantages as a place for outfitting; most of them have a military road leading to the mines, and each one shorter than that of its neighbor – some of them save about one-half the distance. Well, now, that may be well enough, if they can make it win.

---

155 W. G. and J. O. Russell and Valorious Young. See the preceding volume, page 75.

156 A letter from Columbus, Nebraska, written November 22, and appearing in the *Omaha Times* of November 25, 1858.

SANTA FE ITEMS [157]

(The following paragraphs are taken from the *Santa Fe Gazette,* of October 20th.)

PIKE'S PEAK GOLD – We have seen at the store of Messrs. Beck & Johnson a lot of gold dust, some ninety-two dollars in value, which was brought from the newly discovered gold fields, which lie, according to the map, in the region of the head waters of the South Platte and of Arkansas rivers, and in the vicinity of Pike's Peak. . . The specimen of the dust we have seen was brought to New Mexico by Dr. Kavanaugh, who during his recent absence from the territory, traveled over much of the country alluded to, . . .

GOLD MINES NO HUMBUG [158]

Mr. Riethman [159] who went out with the Council

[157] Published in the *Daily Missouri Republican,* November 27, 1858.

[158] *Council Bluffs Bugle,* November 24, 1858.

[159] John J. Riethmann (born in Switzerland, November 28, 1838) became a prominent citizen of Denver. The *Council Bluffs Nonpariel* (copied in the *Missouri Republican* of December 11, 1858) says:

"Mr. Ritchman [Riethmann] says that the miners who have been there long enough to get permanently located and at work, are making, without the aid of long toms or rockers, from two dollars and fifty cents to twenty dollars per day.

"While Mr. Ritchman was there, he saw three dollars and fifty cents worth of gold washed out of a single panful of dirt.

"The largest piece of gold found while he was there that came under his notice, was worth forty-four cents. Old miners have prospected for round gold, and find it in several places, but always in small pieces. He thinks when the snow is melted off in the spring, so that the miners can work in the gulches in the mountains, that round gold will be found in more abundance, and in larger pieces." See A. B. Sanford, "Early-day Experiences of John J. Riethmann," in *The Trail,* xvii, no. 11, pp. 3-6.

Correspondence dated at Council Bluffs, November 29, and published in the *Missouri Republican* of December 10, 1858, says that Mr. Riethmann's return "has caused a perfect jubilee in our city, doubts and despondency no longer finding a resting-place in the minds of any." Some of the letters brought in by Riethmann are published in Section ii of this volume. A. D. Richardson, writing from Leavenworth to the *Boston Journal* (letter pub-

Bluffs train to the South Platte gold mines, returned on the 23rd inst., being only 18 days out from the mines, and brought forty to fifty letters from the boys in the mines, to their friends in this place.

He also brought back about $20 worth of the "dust." He arrived at the mines on the 30th of October, and left on the 5th of November.

He returns for the purpose of aiding his father to go out, and intends to start back in a few days.

PIKE'S PEAK GOLD [160]

We saw yesterday in the possession of Mr. Mead, firm of Messrs. Edward Mead & Co., jewelers, Main street, a quantity of gold which had been brought in from Pike's Peak region. The whole weighed between five and six ounces, was a coarse dust among which was a nugget about the size of a grape shot, and worth over ten dollars. This gold may be seen at the store of Mr. Mead. It was gathered by Mr. McCullough, of Kansas City, and is the largest parcel which has yet appeared here. It will no doubt be followed by others.

THE LATEST FROM THE PLAINS [161]

Lone Skin, N.T., Nov. 14, 1858

MESSRS. PFOUTS & CUNDIFF: GENTS: Trains are still passing here, their stock looking well. . .

There has been no news of interest from the gold mines, although the fever is on the increase. There have been a great many arrivals of emigrants lately at the

lished in the issue of December 10, 1858), says that a gentleman from Lawrence is about to start for Texas to purchase $10,000 worth of oxen for the outfitting towns of eastern Kansas. Others are buying thousands of bushels of corn to supply gold seekers.

[160] *Daily Missouri Republican*, December 4, 1858. This item was copied in the *Chicago Press and Tribune* of December 8.

[161] *St. Joseph Gazette*, December 5; copied in the *Missouri Republican*, December 13, 1858.

mines, and numbers still on the road, and the general opinion now prevails throughout the country, that it will pay well in the spring.

In accordance with this opinion I have started an expedition to the mines under the charge of Mr. Jos. A. Boldoc, who will make a thorough and effectual examination of the country, when I shall be able to give a more reliable account. The young man has had experience in the California mines, and the utmost confidence can be placed in any statements he may make. G. P. BEAUVAIS.

RUSSELL BROTHERS RETURN FROM GOLD REGION [162]

Messrs. Wm. G. and J. O. Russell and V. W. Young, arrived in this city on yesterday, direct from Pike's Peak. They came by the southern route, in twenty-one days, having left the diggings on the 15th of October. The reports from the mines corroborate those we published last week from French's party.

The Messrs. Russell left the Georgia mines on the 9th of last February, with a view of prospecting in the region of Pike's Peak. They arrived at their destination on the 23d of June last, and were the first to put a shovel in the sand on Cherry creek.

Since June, they have been engaged in mining and prospecting, with uniform success. They are old miners, having worked for years, both in California and Georgia, and being thoroughly conversant with mining operations, their opinion in regard to the Cherry creek diggings, is entitled to much consideration. With the present rude implements at the mines, from three to ten dollars can be easily made per day; when proper machinery is introduced, the amount will be increased four fold. . .

---

[162] *Daily Missouri Republican*, December 7, 1858.

The Messrs. Russell left the city yesterday on their way to Georgia. They propose returning in the coming spring with a large company. The specimens of gold they bring are very rich.

According to Mr. Russell's opinion, this will be the last news we will receive direct from Pike's Peak until next spring.

Clark Brothers & Co., bankers in this city, purchased four ounces of the dust from Mr. Russell yesterday morning. They speak satisfactorily of its purity.

THE GOLD EXCITEMENT [163]

We have seen a letter from some parties in England, asking information concerning the gold region of Kansas. The fever seems to have spread across the old Atlantic, as well as to embrace all the states in the union. We hear of companies being organized all through the country, who are preparing to try their hands in the new El Dorado as soon as spring opens. We can confidently assure those who propose trying the new region, that the auriferous deposits are a verity – that the mines will richly remunerate all who devote their energies to the work.

LATEST NEWS FROM THE MINES [164]

Leavenworth City, Dec. 22, 1858

Editor Missouri Democrat: – By the Pike's Peak Express, that reached this city yesterday, bringing nine passengers, I gather highly satisfactory news from the "New Kansas El Dorado." Hitherto I have not been disposed to encourage emigration to the Kansas mines, for fear that the anticipation of those who might go would not be realized, and I would be blameable to

---

[163] *Leavenworth Times*, December 11, 1858.
[164] *Missouri Democrat*, December 29, 1858.

some extent. But that there is plenty of gold, there remains now not a particle of doubt. Most of the passengers I am personally well acquainted with, and feel free to say they are entirely truthful, and their statements may be relied upon fully. They return for their families, and to settle up their business, in time to be on the road early in the spring.

The names and addresses of those returned by the express are as follows: Capt. William Smith (in charge of the express), Johnson co., Mo., Capt. I. T. Parkson, St. Louis, Mo., Lucius E. Chaffer, Wyandotte City, K.T., Wm. Hamblin, Johnson county, K.T., Charles Nichols, Lawrence, K.T., Valorous Ashbrook, St. Louis, Mo., Alonzo Greening, Leavenworth county, K.T.

They left the mines on the 20th of November, being out thirty-one days. They say they will start in the spring from Wyandotte, cross the river on the new bridge three miles above Wyandotte City, take the Santa Fe road, and follow it to the crossing of the Arkansas; thence up said river on the north side to Grant's Old Fort; [165] thence by the Fountaine qui Bouville river to the fork of Pike's Peak and Cherry creek road, a distance of 18 miles. Thence due north to the mouth of Cherry creek on the South Platte river. . . In haste, truly yours, M. W. DELAHAY

ARRIVALS FROM CHERRY CREEK [166]
Leavenworth Times, Dec. 22, 1858
Capt. W. SMITH, L. E. Chaffee, and A. Greeney

165 Probably Gantt's trading-post, built in the early 1830's and located about six miles east of present Pueblo, Colorado.

166 *Leavenworth Times,* December 22, 1858. This was copied in the *Chicago Press and Tribune,* January 1, 1859. The *Missouri Republican* of December 29 copies from the Wyandotte City *Commercial Gazette* a report from the mines by L. E. Chaffee, and reproduces a letter of Mr. Buesche written at the mines on November 20, and one by John Buell from "Aurora City," November 10.

arrived in Leavenworth last evening by the *Pike's Peak Express* (which is making regular trips to and from Leavenworth and the mines). They left the gold region Nov. 20. Came by the southern route, which they pronounce to be worthless, and miserable humbug, as compared with the direct and excellent route from Leavenworth by the Republican fork.[167] They earnestly advise all emigrants to take the latter route, as it is immeasurably the nearest, best and only reliable one. Mr. Chaffee worked two pans of dirt on the Platte from which he obtained 33 cents worth of dust. The dirt was taken from bed rock, one hundred and fifty yards from the river. Only a few Spaniards are mining in the locality from whence this company came. They used wooden bowls, wasted half the dirt and yet succeeded in realizing from $2.00 to $5.00 per day per man. Flour is worth $15 a sack at the mines. Bacon, 30 cents per pound. Sugar and coffee 50 cents per pound. Whiskey, $8 per gallon. This company have established a town, Auraria, from which we have a letter. Capt. Wm. Smith intends running an express to the gold mines at regular intervals this winter. The express will start from Leavenworth.

CORRESPONDENCE ABOUT THE GOLD MINES [168]

Willet House, Florence, Dec. 21, 1858

Editor Times: – As the public generally are anxious to have the latest and most reliable information from the gold region of Nebraska, I hasten to give you what we have here by John Thompson and Adam Bigler, who arrived in this place last night direct from Cherry creek. They came in about 35 days, including some de-

---

[167] The immediately preceding article says that this party plans to return by the Arkansas route. This indicates how reports were twisted to favor different outfitting points and routes.

[168] *Omaha Times*, December 30, 1858.

tention by the way. Mr. Thompson is a resident and merchant of De Soto. Mr. Bigler is a resident and an old citizen of this place. If you have not already been advised of their statements, you are at liberty to publish this, if you deem it will be of any interest to your numerous readers. I give you their statements briefly.

Mr. Bigler left Florence on the 23d of September last, one day after the departure of Capt. A. J. Smith's company – took the northern route up the Platte valley, and crossed the Platte river at Fort Kearny, ten days after leaving here – arrived at Lupton's Fort on the 1st of November – found there a few inches of snow – remained at the fort one week, and then moved on to Cherry creek – remained at the mouth of Cherry creek and vicinity about 18 days – had good weather and plenty of grass the greater part of the way. Immense numbers were flocking to the mines, so that trains were seldom out of sight the whole distance. Mr. B. did not prospect any himself, being engaged in building, etc.; but from what he saw, and from the best information he could derive from others, miners were making on an average five dollars per day with pans only. No "toms" or sluices had as yet been brought into use. The gold is generally distributed over the country. Mr. B. has returned for the purpose of taking his family out in the spring.

Mr. Thompson left De Soto on the 23rd of September last, – travelled with cattle, and also took the North Platte route, crossing at Fort Kearny – arrived at Cherry creek about the same time as Mr. Bigler – remained at the mines about 18 days, during which time he prospected up Cherry creek about 25 miles, then crossed over to the Platte, striking it some 16 miles above the mouth of Cherry creek. He also prospected on some of the small mountain streams running into the Platte. The weather became so cold and there being so much snow on the ground, he could not carry his researches far into the mountains. He thinks he washed out himself about one thousand pans of dirt, and in every one found more or less gold, ranging in value from one to twenty-seven cents. He is entirely satisfied that gold abounds all over the country, and that next season will develope new and far richer fields. The fountain head for this vast quantity of float gold must be in the mountains, but these have been but little explored, on account of the snow. Mr. T. intends to return with his family, and a stock of goods and provisions in the spring, as early as possible. . .

Messrs. Thompson and Bigler left Cherry creek on the 18th of November, with cattle. Gen. Larimer's train (or what was left of it,) had arrived three days previously, having had a very difficult time in getting through. Indeed all the trains south of the Platte experienced excessively cold weather, with a great scarcity of grass. The trains on the north side of the Platte went through in nearly one-half the time of those from the south. Met Judge Townsley's train from Harrison county, Iowa, 100 miles this side of Cherry creek, getting along very well. They had tolerably good weather in returning, until they arrived at O'Fallon's bluffs, when some snow fell, and it continued very cold for a few days or until they arrived at Kearny. The weather then became more pleasant. They crossed at Fort Kearny on the ice. . . Yours, etc., GEO. F. KENEDY.

[ONE THOUSAND DOLLARS A WEEK] [169]
General Eastin has handed us the following letter, for the reliability of which he stands responsible. If

---

[169] The *Leavenworth Times,* copied in the *Lawrence Republican* of January 20, 1859.

true, it makes the gold district a perfect bed of gold.
We give the letter without further comment:

December 29th, 1858

Dear General: I arrived at Leavenworth City yesterday, and as everybody is contributing to your paper something from Cherry creek gold diggings, I thought I would tell you what I did while I was there.

I arrived on the 20th of August, and prospected through the country for some time, and at last struck a vein about twenty-nine and a half miles south of Cherry creek that in richness exceeded anything that had been discovered. I dug out of this vein, in twenty-five days, $3,000 worth of gold, and the vein did not seem to be exhausted in any degree. Yours truly, JOHN HARTMAN [170]

MAJOR VASQUES – GOLD MINES [171]

Major [Louis] Vasques arrived in town yesterday, having come in from his trading post in the mountains. He passed in the vicinity of Cherry creek and other mining localities. He reports to us that the miners are all doing well – making more money than they expected to make at this season of the year. He was in conversation with many who gave accounts as favorable as any we have as yet published. . .

Maj. Vasques is too well known by the business men of the border, the Missouri valley and the merchants of St. Louis, to need any indorsement from us. He has made the mountains and the plains his home for nearly forty years.

There is where he has grown gray, made his fortune

---

170 This appears to be a pure fabrication.

171 Kansas City *Journal of Commerce*, December 30, 1858. For data on Major Vasquez, fur-trader, see L. R. Hafen, "Mountain Men—Louis Vasquez," in *Colorado Magazine*, X, 14-22.

and acquired a reputation among all who know him, to be one of the best informed men upon mountain commerce and mountain affairs now living.

We regret that we are unable to publish more details about the gold, coming from a man of his reputation, whose name mentioned in connection with the country west of us, is the best authority in commercial circles, and among government officials.

WAGONS FOR THE MINES [172]

Our friend, Mr. Geo. Churchill, informs us that he already has quite a number of orders for wagons from companies preparing for the gold mines. He has twelve hands constantly at work, and is about enlarging his shop to meet the demands made on him. He expects to have a full supply of wagons ready for the spring emigration to the gold mines.

LATEST INTELLIGENCE FROM THE GOLD MINES OF WESTERN KANSAS [173]

Just as we were going to press last week, we made the brief announcement of the arrival of Col. Nichols [174] directly from the gold mines. We have since obtained a fuller statement from him.

First as to the gold itself. He confirms the previous accounts of its existence in paying quantities. He says it is scattered more or less over an extent of country 300 miles north and south, and 100 east and west. Over this extent of territory, the colonel prospected more

---

[172] *Lawrence Republican,* December 30, 1858.

[173] *Ibid.,* December 30, 1858.

[174] Charles Nichols, of the original Lawrence party of prospectors, was one of the founders of St. Charles, predecessor of Denver. See data in the preceding volume of this series regarding his town-founding activity. J. C. Smiley (*History of Denver,* 215) says that Nichols returned to his native state (Ohio), enlisted in the 183d regiment of Ohio volunteers as captain, and died in a hospital at Clifton, Tennessee in January 1865.

or less himself, and did not fail of finding gold. The best prospect he found anywhere, was on the Platte, in the South Park, some 80 miles up the river from Cherry creek, and there he is confident the richest diggings will yet be found. This park lies mainly north and west of Pike's Peak. He also found good prospecting at the crossing of the Arkansas at Pueblo.

When the colonel left Cherry creek on the 20th of November, there were from 1000 to 1500 persons in the mines. The newcomers were mainly prospecting, while the old companies were at work building houses for themselves and for their friends they were expecting from the east. There is plenty of cottonwood and similar timber in the mines, and plenty of pine a few miles distant. The miners will pass the winter comfortably in the mines. Among all who had come into the mines this fall, no one who had prospected for a single day had been dissatisfied or discouraged. A company from Doniphan arrived in the evening – a snow storm came up and the next day, without even digging a spade full of earth, the most of them started back for home, cursing the mines and pronouncing them a humbug. They will probably give unfavorable accounts on their return. . .

The New Towns

The spirit of speculation is rife there, as might have been expected from the previous experience that most of the parties have had. Montana is the town laid out by the Lawrence boys, and is on the South Platte, about nine miles above the mouth of Cherry creek. It is a fine settlement, and contains already some thirty houses. At the mouth of Cherry creek there are three towns, viz: Auraria, situated at the mouth of Cherry creek, on its west bank – settled mainly by the Georgians, the Mexicans, and a few men from Nebraska; St. Charles, situated on the east bank of Cherry creek, and is the county seat established by the corps of officers sent out by Gov. Denver; Denver City, so called, is merely a part of St. Charles. These are all the towns that are laid out, although there are people settled along down the Platte for fifty miles.

Squatter Sovereignty in the Mines

The miners are resolved upon carrying out the doctrines of "squatter sovereignty" to the utmost extent. They repudiate the officers sent out by Gov. Denver, and claim the right to select officers for themselves. The Georgia men are the most excited about the matter, and threaten to hang the Denver officers if they attempt to exercise any official functions. The Lawrence boys, being some eight or nine miles distant, look on and laugh at the contest.

The Congressional and Territorial Delegates

The election of a reputed delegate to congress, and another to the Kansas territorial legislature, was a mere partizan affair, gotten up by a set of Buchanan men from Nebraska. The mass of the miners took no part in it. Graham, the congressional delegate, so called, and Smith, the legislative member, are both Buchaneers, and both from Nebraska. They had scarcely arrived in the mines before they got up this sham election, and with true democratic pertinacity are now claiming office.[175] The general feeling among the settlers was, that it was premature to agitate the question of a new territory. The general spirit seems to be, that for the present they can take care of themselves. Indeed, down to the time when Col. Nichols left, there had been no difficulties calling for the interference of law.

---

[175] Graham went to Washington, where he presented a petition asking for a territorial government.

The Journey In

On the 2d [20th] of November, the colonel, with eight companions, left the mines. They came by Smith's Pike's Peak Express, each paying $125 passage money. The journey was made in thirty days. They suffered a good deal from the cold, one of their mules freezing to death. They came by the northern route. As to the vexed question of routes, the colonel thinks there is but little to choose between the northern and southern, but is confident that the middle route, up the Kansas valley, will eventually be found to be the shortest and best. . .

ARRIVALS FROM THE GOLD DISTRICT [176]

Mr. John Elliot, of St. Louis, Mr. William A. Green, of Chicago, and Robert Middleton, arrived in this city last evening [December 31], direct from Pike's Peak, having accomplished the journey in seventeen days – the quickest time ever made upon the plains.

The point of their departure was the old Pueblo Fort, on the Arkansas, 100 miles south of Cherry creek. At this place there were about 200 miners who had come there for winter quarters, bringing their stock.

About three days previous to leaving Pueblo, a party came in from the Platte, who reported nothing doing, in consequence of the severity of the weather. Five days previous to leaving Pueblo, there were several hundred sheep frozen to death in that vicinity. . .

Messrs. Elliot and Green left for the mines in October last, going directly to Pueblo. They were not encouraged by the reports at this point, and concluding that the winter would be severe, they returned. They purpose returning again in the spring. . .

[176] *Leavenworth Times,* January 1, 1859.

FRESH NEWS FROM THE MINES [177]

Mr. O'Donnall, our special correspondent at the gold mines, arrived direct from the South Platte, last Monday evening, [January 10] in fine health and spirits, and bringing the most favorable intelligence in regard to the mines and their future prospects. We hasten to lay his statement before our readers, who will be anxious to learn the facts from one who has made the journey on purpose to find out the truth.

Mr. O'Donnall left Lawrence on the morning of the 23d of September last, for Cherry creek. He took the Arkansas route, and made the journey in forty-four days, including stoppages, one of four days, and another of ten. A part of the company with which Mr. O'Donnall went out, consisted of the democratic officials. Some of these worthies, especially Judge Smith, treated Mr. O'Donnall with a good deal of impudence on learning that he was a correspondent of an "abolition" paper, showing their border ruffian breeding on all occasions. On arriving at Pueblo, the mass of the party, under the persuasions of Smith, Rogers & Co., determined to remain there until spring, not daring to attempt the passage of the divide, which was then white with snow. But by the persuasion of Mr. O'Donnall, by offering to pay one of the company for every head of cattle he should lose in making the attempt, he crossed immediately with a few companions, arriving at Cherry creek about the 3d of November. The rest of the party remained at Pueblo about a week, until the arrival of

---

[177] *Lawrence Republican,* January 13, 1859. In its issue of January 20, 1859, the *Kansas Tribune* tells of the washing at Topeka of some sand brought from the mines by Mr. O'Donnall. "Less than two quarts of the dirt or sand produced about four cents of gold. An old miner washed this out, and says that such diggings would pay from $15 to $20 per day, to the man, when worked with a rocker."

General Larimer's company from Leavenworth, which they joined and came through with. . .

The squatters universally repudiate these officials, and will not recognize them in any manner whatever. Mr. O'D. thinks the miners generally acquiesce in the appointment of Mr. Graham as a delegate to the territorial legislature, and Mr. Smith as delegate to congress.[178] Mr. O'Donnall was in the mines about four weeks. While he was there, the most of the miners were building houses for the winter, mining operations being mainly suspended. Every day, however, parties would be out prospecting and staking off claims, and would come in with from one to three, four or five dollars each, in their pans. All were well satisfied, and looking forward to spring operations in high hope, when they shall be able to construct sluices and ditches for working the gold. . .

The journey in was made by Mr. O'Donnall in thirty-eight days, by way of the South Platte. . .

Mr. O'Donnall brings back some seventy dollars worth of the gold dust as a specimen. A part of it he dug himself, and obtained the rest of miners. Some of the flakes are quite large, one piece being of the value of 67 cents by weight. The whole of his specimens are coarser than what we have previously examined. The gold will be left in the REPUBLICAN office for a few days, where any one desiring can see it. . .

---

[178] Graham went to congress and Smith to the Kansas legislature.

LETTERS FROM THE MINES, WINTER OF 1858-1859

LETTERS FROM THE MINES,
WINTER OF 1858-1859

[LETTER OF DANIEL KNIGHT] [179]

Cherry creek, Oct. 29, 1858

Dear Father: I write you these lines to inform you that after 37 days tedious traveling, we have reached the mines in good health, though the oxen were all lame, etc.

We have plenty of antelope and deer, and there are sheep in the mountains. We have not prospected any yet; but reports are good. Six men made $1300 in two weeks, but the prospect is not very good this winter on account of water, for the stream the gold is on is mostly dry, and we have had to haul the dirt five miles. The average is from $25 to $50 per day on the bank of the Platte now, but there is no telling how long it will last, for the snow is falling fast in the mountains, within fifteen miles of the mines.

Since I commenced writing, a man by the name of Smith, an old mountaineer, who was sent from Fort Kearny to prospect the mines, came into our camp.[180] He says the mines will pay as high as $10 to $25 per day next spring, but there will not be much done this winter, on account of the water being scarce. Court-

---

[179] Daniel Knight was from Sarpy county, Nebraska. This letter was published in the *Omaha Times*, December 23, 1858. Knight became one of the stockholders of the Denver City Town Company.

[180] This was S. W. ("Kearny") Smith, who, with D. C. Oakes, supplied the guide to accompany the Luke Tierney guidebook, reprinted in the preceding volume of this *Southwest Historical Series*.

wright, another old miner, reports diggings on Dry creek that will pay 50 cents to the pan; the diggings are five miles above Cherry creek, and five miles from the Platte river, and there is no water nearer than the Platte. In the spring water comes into Dry creek. He says they are the best diggings found yet. . .

Oct. 31.

P.S. – Since I commenced this letter it has snowed twelve inches in the valley but it is melting fast.

One of the company, John Graves, the same man you bought the apple trees from, has returned from a prospecting tour. He washed about 40 cents out of three pans – the first that he ever washed. Smith, the old mountaineer, is going to Fort Kearny this winter, and will probably be in Omaha. If he is there, inquire of him, and he will tell you the truth; you will hear all about it from him. My ink is frozen and I have to finish with a pencil.

Direct one letter to Cherry creek, by way of Fort Laramie, and one by way of Fort Kearny. DANIEL KNIGHT.

[LETTER OF E. P. STOUT] [181]

Cherry creek, Oct., 30th, 1858

. . . Our company, consisting of Forbes, Jackson and myself from De Soto; A. J. Smith, of Florence, with

[181] Mr. Stout's letter was published in the *Omaha Nebraskian,* and copied in the *Missouri Republican* of December 22, 1858. As an introduction to the letter the newspaper says: "Mr. Stout went to the gold mines from De Soto, county seat of Washington county [Nebraska], in September last. Mr. Stout is too well and favorably known in Nebraska to require any vouchers for his respectability and veracity." Stout became the first president of the Denver City Town Company, upon its organization November 22, 1858. A street of Denver is named for him. "The Contract with Specifications for Building Stout's Pioneer Cabin in Denver, 1858," was published in *Colorado Magazine,* IX, 169-170. His photograph appears in Smiley's *History of Denver,* 214. In that volume (page 193) Stout is quoted as saying: "Our party, consisting of myself and two others, with a four-horse team, left

ELISHA P. STOUT
First president of the Denver City Town Company

three teams, arrived here on the 26th. Found the Platts-mouth and Iowa Point company in, a few days ahead of us, but no company in from north of Platte. There are, probably, not to exceed five hundred persons in here yet. The Council Bluffs and Omaha company will be here to-morrow. . . There has been but little steady mining done as yet. All seem to be prospecting for some-thing big, and are not satisfied with a reasonable com-pensation. The few that are mining regularly take out from $3 to $5 per day, and travel five miles morning and night to and from camp.

Sixteen dollars per hand has been taken out by Green Russell's party of Georgians. He has now gone back to bring a party of three hundred to drain Cherry creek in or by spring. They propose draining twenty-seven miles, so as to get to the bed of the stream. It will un-doubtedly be rich.

There has no gold been found as yet except the float gold. The round gold or quartz will probably be dis-covered along the foot or up in the mountains. There are six or eight of the Plattsmouth company, and a com-pany of Spaniards up there now prospecting, and will be back in three or four days. 'Tis too cold to do much prospecting in the mountains this winter; they are now covered with snow. The Black hills are about eight miles from here and principal range of the Rocky mountains, about thirty miles. Pike's Peak is eighty miles distant and in plain view. Russell's party spent a couple of months there, and they say there is no gold at that point.

Omaha on the 26th day of September, 1858, and arrived at the mouth of Cherry creek on the 26th day of October." Smiley says (page 213) that Stout was then (1903) living in Wyoming (a suburb of Cincinnati) and was presi-dent of the Cincinnati Savings Society.

[D. F. RICHARDS] [182]
Winter quarters on Platte, 2 miles from mouth
Cherry creek, Oct. 30th, 1858

S. A. Megeath, Esq. Dear Sir: – We arrived today all
well and hearty. Our cattle stood it tolerable well considering our loss by a stampede which occurred about
40 miles this side of Fort Kearny, by which we lost one
yoke of cattle. But we were fortunate enough to obtain another yoke at the fort, which enabled us to get
through. We are two miles east of Cherry creek, five
miles from the foot of the mountains, and have a good
view of Pike's Peak which is 70 miles from the mouth
of Cherry creek. We saw a great many Indians, both
Cheyennes and Sioux, but experienced no difficulty
with them more than having to drive late sometimes to
get away from them to camp.

Concerning the gold mines I can say but very little
of what I have seen myself, for I have not had time
yet to prospect any, but I will relate what I have reason to believe, from what I am told by persons who have
been here long enough to look around some. One of our
party whom we sent in advance of the train to secure
our quarters, and he, in company with Kearny Smith,
prospected thirteen days before we arrived, and both
tell about the same thing concerning the chance for
gold. I saw, myself, a pan washed by Mr. Grave's party,
who got here a day before us, from which the dirt
yielded 20 or 25 small scales; there was but a shovel
full of dirt, and it was taken from the Platte bottom,
just where we are camping. All the sand which I have
seen that contains gold, strongly resembles the black
drying sand used for drying ink. . .

Some Mexicans packed over some flour and could

182 *Omaha Times*, December 16, 1858.

not sell it for $12 per hundred, so they took it down to trade with the Indians; but flour and groceries will be in demand in the spring at good prices. The distance the route we came, up the South fork of the Platte, is 570 miles. The distance from here to Fort Laramie is about 170 miles, so that would make it 670 miles by that route. There will be plenty of feed in the spring. Yours, etc., D. F. RICHARDS.

[EXTRACT FROM LETTER OF JOHN GRAVES] [183]

We found the earth pretty much pregnated with gold more or less. We can find the "color" in most every place, but, as a general thing, the soil would not pay unless extensively worked. But up Cherry creek and other branches, it will pay in almost any way you choose to work it. We will not, however, be able to do much this winter, but those who winter here will have the advantage in the spring, while the water is coming down the mountains. I think we shall be able to make from three to ten dollars per day, but hope to do better with the implements we have to work with.

I send you a small particle of "dust," to show you that the gold is not all a humbug. This I panned out myself. . .

Emigrants are daily arriving, and we think there will be from four to six hundred men in here this winter. We hear and fear but little of the Indians. There is a town laid off here, but as yet no splendid brick blocks have gone up.

Snow has been falling all the morning, and it is now about seven inches deep, and yet falling.

---

[183] *Missouri Republican,* December 11, 1858, copying from the *Council Bluffs Nonpareil.* John Graves was one of the original stockholders of the Auraria Town Company (predecessor of Denver). See the original record in the library of the State Historical Society of Colorado.

[W. R. REED] [184]

Winter quarters, 2½ miles from Cherry creek.

75 miles from Pike's Peak, Nov. 1, 1858

Friends: We arrived here on the 30th of October, and during the night there was a heavy snowstorm – snow fell about a foot deep – this gave the boys the blues. They will have their houses completed tomorrow. Knight and S. Dillin has split, and I am with Jack the butcher. He has been here about ten days – he sends his best respects to you. I have had bad luck on this trip – lost my cattle and broke my shoulder all to d—d smash, running after buffalo. I went back to Fort Kearny and traded my pony for cattle, got my shoulder set and started back. The boys agreed to wait for me forty miles above Kearny, but when I got there they had left. John Harsed was with me – he had bought two yoke of cattle. When we arrived at Cottonwood springs, we learned that the boys were three days ahead of us, so we pushed on and caught them 160 miles from Fort Kearny. I gave them a good blessing for leaving us. We had no grub with us and had to beg our way, and two days we were without a bite – don't you think I cussed them? I left S. Dillin and Jack took me in. My shoulder is getting so I can use it a little.

About the gold. About two weeks before I got here, there was some men going to the states, they showed $100 worth of the stuff. The man that carries this letter had $12.50 of the gold. There can't be any mining done this winter. I believe that the gold is here. There are some men making from $1 to $3 a day – that is on the Platte. . . Your friend, W. R. REED.

---

[184] The *Council Bluffs Bugle,* of November 24, 1858, tells of the return from the mines of John J. Riethmann (mentioned above) and says: "He brought us a letter from W. R. Reed, who had been in our employ for over a year before his departure for the mines, which we publish below."

Postscript: John Graves has been prospecting this morning and found 50 cents worth of dust.

Tell Sheriff Baker, that his horse thief James Gilson is out here and goes by the name of Fox.

There is a little town above here, and day before yesterday there was two men killed there. Blake and Williams have not got here yet.

[SAMUEL S. CURTIS] [185]

Winter quarters, 2½ miles below Cherry creek,
Tuesday, Nov. 2, 1858

Dear Henry: Behold us at last at our journey's end, all well and hearty!.. Our shanty is about half finished. We will probably get into it on Thursday. The days are warm and pleasant, but the nights are cold. . .

I was up at the mouth of Cherry creek yesterday, at the organization of the Aurora [Auraria] Town Company. It is going to be the place; one hundred original stockholders are each to build a house. If present prospects are realized in the spring, come out as early as possible. . .

---

[185] This letter was written to Curtis's brother in Omaha. It was printed in the *Omaha Republican* and copied in the *Missouri Democrat* of December 4, 1858. Curtis street, Denver, was named for S. S. Curtis. He was an early postmaster of this city. He joined the Union army in the Civil war, his father (Samuel R. Curtis) being a prominent officer. In his later years Samuel S. Curtis lived in Omaha. He came to Denver in June 1911 to attend the dedication of the Pioneer Monument on the Civic Center. Extracts from a letter of his written from Cherry creek, October 30, 1858, were printed in the *Missouri Republican* of December 11, 1858. They contain no important information. Curtis contributed data for J. C. Smiley's *History of Denver* (1903). On page 202 the following quotation from one of his letters is given: "I left Council Bluffs, Iowa, for 'Pike's Peak' on September 20, 1858, and at Columbus, Nebraska, was elected captain of a train of emigrants. At Plum creek, thirty miles west of Fort Kearny, the train lost thirteen head of cattle by stampede, which threw us behind some other trains; otherwise we would have been the first train of emigrants to reach Cherry creek after the discovery of gold by the Georgia company. We reached camp about one and one-half miles below the creek, on October 30th."

I think we will make something this winter (when it is too cold to mine) sawing lumber with our whip saw, as there will be a great demand for it in the spring, to make longtoms, cradles, sluices, etc.

We are living on antelope and venison. Antelope is ahead of any meat I ever tasted. We are living about twenty miles from the mountains, and they loom up like a steamboat in a fog. I must get to sawing logs to cover the house.

I enclose you some Cherry creek gold, in its native purity. Yours truly, SAM S. CURTIS.

[H. C. ROGERS] [186]
Camp on the South Platte, in the gold regions,
Nov. 2.

Dear Col: We arrived on Cherry creek on the evening of the 30th of last month, and on this stream to-day, where we expect to spend the winter. Some of the boys went out a few moments since with a pan, and in one pan of earth they found about twenty cents worth of gold. I have but little doubt of there being plenty of gold here, but I find the weather too cold for mining. I can't tell much about the country as the ground is covered with snow. We are all well and had a glorious time crossing the plains. Plenty of game; deer, antelope and bear. I have an apportunity to send this by a man who is going to the states. As we have no mail I will watch the chances. Your friend, H. C. ROGERS.

---

186 This letter was published in the Kansas City *Journal of Commerce,* December 12, 1858. In introducing the letter the newspaper says: "The following letter was received yesterday by Col. N. C. Claiborne. The writer, Mr. Rogers, was previous to his departure for the mines, a partner of the colonel in the practice of law. He is a gentleman known in this city, and what he says will meet with an indorsement from the members of his profession in this city, and from all his acquaintances."

[DR. G. N. WOODWARD] [187]

November 2, 1858

Dear Friend: We are now camped on the South Platte, six miles above the mouth of Cherry creek. We arrived here to-day and shall look around for winter quarters tomorrow. The company are all well. There are about three hundred men on the river now, and more expected. I saw what gold was washed from six pans of earth. I should judge it to be worth $2. There is gold here, and in quantities to pay largely in the spring, when we can use sluices. We have had a severe snow storm for the last few days but it is more mild now. There is timber and grass on this river, and we think our animals will winter well.

We have had a long trip from Kansas City here, but came through safe and well. We had no trouble with any one. We do not expect to do much this winter except to get ready for work in the spring.

Men here are making from $3 to $8 per day, with pans.

I send this by a man who starts in to-day, so that I have no time to write more. I remain yours, G. N. WOODWARD.

P.S. I have washed out one panful and got twenty cents worth of gold.

[H. L. BOONE] [188]

South Platte, Nov. 2, 1858

Messrs. Editors: Our company, composed of some sixty

187 *Ibid.*, December 12, 1858.

188 *Ibid.*, December 12, 1858. See footnote 105, above, for biographical data on Boone. Another letter from Auraria on November 2, was published in the *Kansas Tribune* of January 6, 1859. It is unsigned, but undoubtedly written by William McKimens, whose almost identical letter was published as Appendix C in the preceding volume of this series.

men are now in camp on the bank of the South Platte,
some four miles above the mouth of Cherry creek. We
arrived here early this morning, being some forty-five
days out from the states. Our trip, instead of being, as
we anticipated, one of hardships, proved to be rather
a pleasant trip. We were favored with most delightful
weather for the first six weeks of our journey, but un-
fortunately, we were overtaken by a heavy snow storm
on the great divide between the waters of the Fontaine
qui Bouille and Cherry creek, which prevented our
traveling for a day or two. At this point we met with
the remaining portion of the Lawrence company, composed of some eighteen men. They have erected comfortable cabins for winter quarters on this river, and
design laying off a town site, or settlement, built up by
a company of old mountaineers, in anticipation of a
heavy emigration next spring. Above us, also is another
small settlement built up by a company of Iowa men,
they number 19.

And now so far as gold is concerned, I would remark
that we have found gold on all the streams where we
have prospected, on the Fontaine qui Bouille, on
Cherry creek and on the Platte. The question is at
length settled with us, it is a fixed fact that there is gold
here. I do not say that we have found it in abundance,
neither do I believe that it is here in very large quantities, but yet sufficient to pay the laboring man from
$5 to $10 per day. Not more than fifteen rods from
where I now sit, the Lawrence company are now taking
out from $5 to $9 per day.

My impression is that no amount of labor will ever
develop riches, such as we have found in California,
and yet I believe that there is gold sufficient to justify
the thousands of emigrants that will come in next
spring. . .

A very amusing incident occurred here a few minutes ago. As I wrote I suggested to my friend, Mr. Winchester, that I had no money to buy postage stamps for my letter. He remarked that he would go and dig some for me. He washed out a few pans and returned with thirteen cents; and to him, therefore, you are indebted for this letter. . . With respect, yours, HAMP. L. BOON.

[H. MURAT] [189]

Montana. On South Platte, Nov. 5, 1858

Friend Zillhard: On Thursday morning Nov. 3d, we arrived here all in good health. The whole journey was more a pleasure trip than anything else. We had most beautiful and pleasant weather the whole trip, with the exception of two thunder storms, which troubled us somewhat. The last day we ascended to the height of about 6000 feet above Kansas City. My wife is well and getting fat. She looks as blooming and fresh as a maiden, so well has the free air of the prairies agreed with her.

Yesterday myself, Philip, and a young man from Westport, felled the trees for our log cabin in six hours, and to-day we hauled them in with our oxen, and to-morrow we shall have the cabin done ready for occupancy.[190]

My dear friend, we are not sorry for coming out here, for in the first place it is the most lovely country you

---

[189] This letter, published in the Kansas City *Journal of Commerce,* December 14, 1858, is introduced as follows: "This is a letter written in German, to one of our business men. And we are under many obligations to our friend Mr. Bodenweiser for the following translation." "Count" Murat was an interesting character of pioneer Denver.

[190] This cabin, first erected in Montana (near present Overland park, Denver) was later moved to Auraria, where it became a hotel. In it was born Auraria Hubbel, first white girl born in Denver. The old house was taken down in 1938 and moved to the place of Mrs. May Bonfils Berryman, near Denver.

ever saw. To our right there is a range of mountains where the Platte river emerges. It must be a most beautiful sight in the summer. Gold is found everywhere you stick your shovel, paying from five to ten cents to the pan while prospecting, and there is no doubt but what it will pay from $10 to $20 per day to the man.

As I remarked above, gold is here plenty, and as soon as spring makes its appearance, the whole world will be in a blaze of astonishment at the riches that will be taken out of the earth.

For this winter we cannot think of mining, because we have to finish our house for shelter this winter. This done, we shall begin hunting to get fresh meat. Our oxen we want for other purposes, but when spring comes we shall be in for the gold.

Yesterday one hundred and ten men arrived from Nebraska. We count now, all told, 250 men.

Thirty wagons more are expected in every day. My *frau* is the first white woman [191] out here and will make money by washing clothes, which will pay her perhaps fifty cents apiece. You must at any rate come out here, you can make your fortune as a shoe maker. But come soon, because every day you lose will be a pity. I shall send you some gold before you start, and I want you to bring me out some things, such as flour, etc. etc.

The last named article is selling here at fifteen dollars per sack, which is cheap to what it will be soon. . . Yours truly, H. MURAT.

[A. C. EDWARDS] [192]

St. Vrain's Fort, Nov. 6, 1858

Dear Father: – I am well and in good spirits. We are

---

[191] Mrs. Murat, though a prominent pioneer, was not the first white woman in the settlement. That distinction goes to Mrs. Rebecca Rooker.

[192] Published in the Brownville *Nebraska Advertiser* of December 10, 1858. This newspaper says this letter and others were received December

within two days' drive of the mines. The news is of the most flattering character. We are in camp with a returning company by whom I send this letter. They have been in the mines, and explored them thoroughly, and have quantities of the gold dust with them. Miners are making from three to ten dollars per day. There is gold anywhere along here; the only difficulty is to get water with which to wash the dirt.

We are now thirty days out from Brownville. Had a snowstorm on the 29th and 30th of October. The weather here is now fair. . .

Tell John and John J. and Watt Richardson, and others to come on early in the spring. Bring clothing and eatables; flour, sugar, coffee and bacon. Flour is worth $15 per sack; bacon 40 cents per lb; sugar and coffee 50 cts. Bring your mining pans and buckets with you. Start early; don't wait for grass, but haul your feed for your stock.

It's a mistake about 1500 men being in the mines; there is 500 or 800.

I will write you again after I prospect, as soon as I meet with an opportunity of sending. Yours, A. C. EDWARDS.

[R. C. BERGER] [193]

St. Vrain's Fort, Nov. 6, '58.

Dear Brother: To-night we are encamped with a company of miners who are returning for an outfit in the

9, and were from the Brownville company that had set out on October 7. St. Vrain's Fort was near present Platteville. Another letter from Edwards, written from "Mountania," November 17, was published in the *Nebraska Advertiser,* December 23, 1858.

193 *Ibid.,* December 16, 1858. A letter from another member of the same company, Joseph Baker, was published in the same paper. Another of Baker's letters, written from Cherry creek, November 15, was published in the *Nebraska Advertiser* of December 23, 1858. John S. Hardin's letter of November 6, from Fort St. Vrain, was also published in the *Nebraska Ad-*

spring. I have therefore, an opportunity of writing you. . .

There are about 800 men in the mines, and more coming continually. An election will be held next week to elect a representative of the mining interests in the territorial legislature.

Tell Dave Seigle to bring out a stock of his clothing; they will sell like hot cakes. Liquors, groceries, and all kinds of provisions, will be, in fact are, in great demand.

I have not time to write more; my ink is freezing, and I finish this with a pencil.

The bearers hereof will make a full report of their explorations for the Plattsmouth or Pacific City papers, to which you can refer and reply upon.

Tell all the boys to come in the spring. Yours, R. C. BERGER.

[W. W. HOOPES] [194]

Cherry creek, Nov. 7, 1858

Mr. J. W. Pattison: . . We arrived here on the 3rd inst., having had glorious fine weather nearly all the way through. The day before we reached this place, snow fell to the depth of five or six inches; it has, however, nearly all disappeared.

For ten miles above Cherry creek, and eighteen miles below its mouth, the passer-by may hear the heavy stroke of the miners' axes upon the unfortunate cottonwood along the route, and the *wo haw* and *gee wo* of

vertiser, December 23, 1858. In this same paper is another letter from R. C. Berger, dated November 17, 1858. One of his, dated November 27, appeared in the *Nebraska Advertiser* of January 13, 1859. After Berger returned home he wrote a general statement to the public (published in the *Nebraska Advertiser* of March 10, 1859) in which he says that "the gold diggings of Cherry creek have not turned out as was expected by me when I wrote those letters."

[194] *Omaha Times,* December 9, 1858. The first issue of the *Rocky Mountain News* (April 23, 1859), tells of the election of W. W. Hoopes as auditor of Arapahoe county (Kansas) on March 28.

sturdy would-be miners, who are now busy hauling logs and firewood to their winter quarters – all is life and bustle. I do not know how many emigrants have arrived, but I should think between four and five hundred men.

There has been some prospecting done but it is uncertain what the result has been, as those who prospect keep a close mouth on subjects relating to gold. . . Your most ob't serv't, WM. W. HOOPES.

[W. D. McLAIN] [195]
Fifteen miles below the mouth of Cherry creek,
Nov. 7, 1858

Dear Doctor and Family: – I arrived at the mouth of Cherry creek on Wednesday last, the 3rd instant, about noon, and went to work to build a house. I then cut my first tree, and before night I had 23 down, and four blisters on my left hand as large as a dime. That night, whilst we were eating supper, an old trapper came into camp, and we gave him his supper. It pleased him so well that he offered us a house of his which was near the one he occupied, and the next day we bundled up and came down the river fifteen miles, and moved into the house that night. The next morning I went out before breakfast and killed two deer, large ones, and yesterday morning I killed another so we have plenty of fresh meat; we intend using considerable meat, in order to save our flour, as it is worth $20 a sack, and meat can be had for the powder and shot used in the killing of it, and not go far for it. The deer I shot were not 12 rods from the house. There is plenty of elk, deer, antelope,

---

[195] The *Omaha Times,* which published this letter on December 9, 1858, says that McLain was well known in Omaha and could be depended upon. W. D. McLain became captain of a battery of field artillery organized in Colorado in 1862. His name is on the original list of stockholders of the Auraria Town Company.

turkeys, and some bears, but I have not seen any yet, and am not anxious to.

We made the trip in 37 days, had good luck, and were in a train of 21 wagons after we left Ft. Kearny. There were 67 men, 106 cattle, and 10 horses, and of all the lot there was but one ox died; there was no sickness in the train, which was very strange for we had three doctors along. . . And now for the main question, is there gold here? there is and plenty of it. It is too late now to dig as the ground is frozen and cannot be washed; but there are men here who were here in time to mine some before the cold weather set in, and they made from $5 to $10 per day with a pan or cradle. It averages ten cents to the pan, and they say a man can double or treble the amount per day with a long tom or sluice. I have seen plenty of the gold so I know it is here. We have taken the first steps toward organizing a territory, sending a delegate, etc.[196] I guessed at the distance as I came along and made it 597 miles. Yours etc., W. D. McLain.

[E. C. Mather] [197]
Camp on Platte river, 2 miles from Cherry creek,
Nov. 7, 1858

This, dear friend, is a busy time. We arrived in safety at this point all well, on Monday a week ago; the incidents of the trip I cannot give you at present, as I have but a few moments to write. The first thing we did was

---

[196] At the election held on November 6, Hiram J. Graham was elected a delegate to congress and A. J. Smith a representative to the Kansas legislature. See Milo Fellows, "The First Congressional Election in Colorado (1858)," in *Colorado Magazine*, VI, 46-47.

[197] *Omaha Times*, December 9, 1858. Letters of the same general tone, written from the mines on the same date (November 7), were published in the Kansas City *Journal of Commerce* of January 4 (letter signed L.B.R.) and January 15 (letter of Robert B. Willis).

building our house, which we have completed. To-day Spooner and myself went up to Cherry creek, there are two towns laid out at the mouth of the creek, one on either side, it will be a good point I have no doubt.

Now something about the gold mines; that it is a fixed fact there cannot be any further doubt – that the mines are immensely rich is also a fact...

I would advise you to come out in the spring as early as possible, and be sure and bring a plow and a supply of seeds, bring a light plow, as the soil is very loose and sandy, it is well adapted to raising wheat and corn; there can be fortunes made in farming.

There is a large amount of provisions brought from New Mexico to this place. Mexican flour is only $10 per sack; it is not as good as our S.F. flour, but is good. Sugar, tea, coffee and fruits, you want to bring with you.

Write to me immediately, direct to Fort Laramie, as we shall have an express running from there to this place during the winter. I will write you again soon when I hope I shall have more leisure time, at present I am hurried. We have done no mining yet but shall soon. Yours truly, E. C. MATHER.

[JOSIAH HINMAN] [198]

Montana, South Platte, Nov. 8, 1858.

Dear Parents: – I have not had an opportunity to write you for two months or more. I suppose you are all anxious to know something reliable in reference to this country; there is gold here, but our means of get-

---

[198] Letter published in the *Beloit* (Wisconsin) *Journal* and copied in the Kansas City *Journal of Commerce*, January 19, 1859. Hinman, one of the original Lawrence party, had recently come to Kansas from Wisconsin. He was president of the company that founded the town of Montana. See Smiley, *History of Denver*, 190.

ting it are limited; some of our men however are making as high as ten dollars per day. . . . There are now on the South Platte over three hundred men, and they inform us that there are at least three hundred more on the road. So much for the country and things in general. As for myself, I am well, my food consists of venison, antelope, bear, and elk meat and bread, and as long as I can get flour I won't ask for anything better. I cannot tell you when I shall come home, perhaps in a year; tell Moses to come out here by all means, that is, unless he is making money very fast where he is. . .

I am interested in every town in this country, but I think Montana will make the most important place. . .

The snow fell the other day about a foot deep and I expect we shall have a bad winter, but I am well clothed in buckskins and furs and shall live as comfortably as Robinson Crusoe ever did. I never was more happy and contented in my life; the Indians have sent us word that we must leave this country, however, we don't fear them much. I will write you again as soon as possible, but I shall probably not have the chance before spring.

My love to all the family, and regards to all inquiring friends. Very affectionately, JOSIAH HINMAN.
P.S. – Direct your letters and papers as follows: J. Hinman, Montana, Arrapahoe county, Kansas territory, Fort Laramie post office.

[B. BARNHISIL] [199]

Near Cherry creek, Nov. 8, 1858

Editor of the Times: – At last I have safely arrived at this noted place. We have taken up our quarters on an island, thirteen miles below Cherry creek, on the

---

[199] Published in the *Leavenworth Times,* December 18, 1858.

South Platte. We are comfortably fixed; have a good cabin that was built some years ago by some trader, and we have the use of it until spring.[200] There is also plenty of timber and good grass to winter our mules, and so we are better fixed than any that are on the Platte at this time. Ours is the only cabin on the Platte that is finished, although there [are] many under way at this time. We arrived here on the 15th of October, just thirty days from Leavenworth. I enjoyed the trip very much, and I am contented so far. I have been doing nothing of any account but hunt. We can't mine until spring on account of water and cold weather. . .

In regard to the gold in this country, I can't say much. I was up in the gold regions about four days prospecting. We found gold, but in small grains; whether it will pay to any extent I cannot say until we get fixed to work. There will be good diggings found next summer. . .

The gold news has caused a devil of excitement in Iowa and Nebraska. When we came up the Platte we were the second team, and since we got here, there have at least one hundred teams passed this island, all from Nebraska and Iowa, and they are still coming. A train of twenty-five wagons came on the southern route – report plenty more on the way.

If there is gold in great quantities it will be found by next May, and if it pays, then is the time to come. There is no doubt about the gold, but the question is whether it will pay. All the gold is float gold, very fine, and is only found immediately on the bed rock. Some

---

[200] John S. Smith, Elbridge Gerry, James Sanders, and several other mountain men had been trading with the Indians in this region for years. This is one of the few references to the existence of a log cabin in this vicinity before the gold rush. It was probably Sander's ranch, on what was later known as Henderson's island.

places you can find the bed rock by digging three feet, and some places twenty feet, and often not at all. Yours truly, B. BARNHISIL.

―――――

[JOHN BUELL] [201]
Aurora [Auraria] City, November 10, 1858
Our cabin is completed to-day. Kennedy and myself have done all the work. We have been amazingly industrious. The city from which I write, is a new but prosperous burg. The lands about this city are splendid for farming purposes. The soil produces everything, and the climate is luxurious. Opposite my cabin, miners are making from five to eight dollars a man per day. They carry the dirt some twenty yards to the river. . .

It is the general opinion hereabouts, by those well qualified to judge, that this South Park region, at the headwaters of the Rio Grande, Colorado, Arkansas, and South Platte rivers, will produce a new El Dorado.

The other prominent point in the gold region is on the Platte river, to the north of Pike's Peak, on the old town site of El Paso.[202] At both these points sawmills are to be established. The South Park is at present only accessible to those who have pack mules, and, consequently, El Paso will have the greatest temporary success, though, ultimately the great bulk of miners will force their way to the point where the diggings

―――――

201 *Leavenworth Times,* December 22, 1858. A similar letter written from "Platte valley, November 18," and signed "John L. Buell," was published in the *Leavenworth Ledger* of December 22, 1858. "John J. Buel" (probably the same person) helped plat the sites of Denver, Boulder, and Central City, Colorado; he served in the Civil war and was the custodian of Mason and Slidell, Confederate emissaries to England; in later years he settled in Michigan and founded Quinnesec, where he died in 1917, at the age of 81, according to *The Trail,* IX, no. 10, p. 29.

202 This was at or near the present site of Colorado Springs. The projected town was named for Ute pass, leading to South Park. Buell's location is in error.

are most profitable. The Georgia company were deterred from exploring the South Park, in consequence of hostile Indians, who would not permit exploration to be made. There are at present from four to five hundred miners at work. Most of them are comfortably housed in winter quarters. They work at intervals. The weather, as yet, has not been so severe as to interfere with mining operations.

Emigrants are arriving daily by scores. We have an abundance of provisions, and the health of the district is unexceptionable. Yours truly, JOHN BUELL.

[F. W. RIGG] [203]

Cherry creek, Nov. 10, 1858

Mr. J. C. Evans: Dear Sir: . . . I have been here about fourteen days, and I have prospected the greater portion of the time, and find gold in many places in sufficient quantities to pay from two to three dollars per day to the man with the shovel pick and pan, and according to that, three men, with a good long tom, could safely count on from $18 to $24 per day. I have taken out in all about $16; but as I stated before, the greater part of my time has been spent in prospecting. We have concluded to spend the winter here, and are now preparing winter quarters. . . I say come early in the spring, and come by the Kansas City route; that is the route by which I came, and I have talked with various persons who came by the Leavenworth route, and I find the southern route much preferable to any other, and more especially so early in the spring as you will wish to start; as grass comes on this route at least two weeks earlier than on any other.

---

[203] Published in the Kansas City *Journal of Commerce,* December 14, 1858. Mr. Rigg was from Kansas City.

I will send you this by hand, and as the company are about starting I must close. Respectfully your friend, F. W. RIGG.

[W. O'DONNALL] [204]

Montana, Nov. 18, 1858

On the 14th of this month our train arrived at the gold regions, and here we found everything even more encouraging than we anticipated. That gold exists here, and plenty of it, is now placed beyond a doubt. I found some twenty-five men busy digging out the gold, and making from four to seven dollars per day with their pans. By introducing the common rocker, they can make three times as much as they can with the pans. We found no snow except a few inches, which causes no hindrance. We shall winter our cattle in the valley adjoining Montana. This is a town the Lawrence boys laid out. There are now thirty-four cabins on the site; the inhabitants will number 150 – and only one woman [Mrs. Murat]. . .

[204] Published in the *Lawrence Republican,* December 30, 1858, and written by that paper's "special correspondent at the mines." A letter written by "C.F.C.," from "Three miles below mouth of Cherry creek, November 17," was published in the *Council Bluffs Bugle* of December 29, 1858. It contains no new or special information. The *Omaha Times* of January 6, 1859 quotes the following extract from letters of A. J. Williams, written November 17 and 20, and published in the *Crescent City Oracle* of December 31, 1858: "I believe we are in the midst of a gold field of unknown extent and of fabulous richness—the precious ore can be found anywhere—everywhere on reaching the rock bed, and is generally scattered through the sand, clay and soil. There is a broad, fertile plain or bottomland some twenty miles wide and goodness knows how long—good land for agriculture or to wash gold from. There are twenty-five houses already up in town, and 163 under construction. There are large bodies of pine timber a little way out in the mountains. Clancy is here, building and knocking about industriously. Landers [Sanders] & Co. have established a monthly express between here and Ft. Laramie, through which we receive our mail and other matter. William Smith also started an express from this place to Leavenworth. Old mountaineers say that the climate here is similar to that of New Mexico. The weather thus far has been exceedingly fine, and some are out washing for gold, but most are building houses and preparing for winter."

The officers sent out by Gov. Denver, won't "go down" with the people here.[205] It appears that Gov. Denver furnished the man [H.P.A.] Smith with blank commissions, ready for him to fill out in case he found men of the right stamp. He had no reputation of his own, except that of being a lacquey of his excellency. He met Gen. [William] Larimer, of Leavenworth, between Puebla [Pueblo] and this place, and induced him to accept the treasurership of this county; but all their figuring won't "go down." Larimer is already ashamed of what he has done. He told me that he was chairman of a Republican convention a few days before he left Leavenworth. . . W. O'DONNALL

[M. M. JEWETT] [206]
Auraria, Nov. 18, 1858

We arrived here on the evening of the 16, after a very pleasant journey. This place is situated on the junction of the Platte and Cherry creek, and about 12 miles from the mountains. I went up to Montana, a place of about 30 log houses, six miles above here, on the Platte. Found Buell and company – also most of the Lawrence boys. Whitsett and I are going to build at Montana, Lawrence and Dorset at Auraria. It is estimated there are almost six hundred men within fifteen miles of this place, scattered along the river.

As I am lying on the ground, writing on a saddle-skirt, by the light of our campfire, I have not time to write in detail. Mr. Greening, of Salt Lake valley, will bring you this. He starts early in the morning.

205 These were officers for Arapahoe county, Kansas.

206 *Leavenworth Times,* December 25, 1858. Jewett was a member of the "Leavenworth party," of which Larimer, Richard E. Whitsett, C. A. Lawrence, and Folsom Dorsett were also members. A less interesting letter, written by Larimer from "Denver City, November 19," also was published in the same newspaper. This appears to be the first newspaper publication of a letter written under the heading "Denver City."

If I had a small stock of well selected goods with me, I could coin money this winter. Whiskey is $8 per gal.; coffee and sugar, 50c per lb.; gun caps, 50c per box; and everything else in the same proportion. We have every assurance that there is gold here. Mr. Turner, of the Lawrence party, thinks that with water in the spring one can make from $10 to $20 per day.

Those who are washing now are making from $2 to $5 per day. . .

We will have an express from Laramie every month. Yours truly, M. M. JEWETT.

[G. N. WOODWARD] [207]

South Platte, November 19, 1858

Messrs. Van Horn and Abeel: You have no doubt heard reliable news of the gold produced in this vicinity. There has not been any apparatus for washing – except an old apology for a rocker. The men now here are making arrangements for sluices and long toms, to use in the spring. We find from washing in pans, from five to fifty cents per pan full. The gold is found from the South Platte as far south as the Arkansas, and 200 miles into the mountains.

This company have located and laid out a town or city, at the junction of Plumb creek with the South Platte, and named it Pike's Peak City.[208] It is about ten miles from the mountains, and in the center of the best mining region at present, and is surrounded by the best of pine timber. If we only had a saw mill, with the fine water power of Plumb creek, a man could

---

[207] Published in the Kansas City *Journal of Commerce*, December 25, 1858. John G. Harris wrote from Auraria, on November 19, a letter that was published in the *Missouri Republican* of January 12, 1859. This paper speaks of Harris as "a well known Missouri river pilot, who was lately seized with the gold fever." Harris writes favorably of the mines.

[208] The town never materialized.

make a fortune – there is no pine timber below Plumb creek, on the South Platte. We intend building on our new location this winter. All the trade from the mountain miners must be done at this point, and if your friends wish to bring out goods in the spring, that is the point to sell them at. We are now camped for winter quarters 6 miles about the mouth of Cherry creek. The best route for emigrants is to Fort Riley, thence up the Smoky Hill fork, which rises within thirty miles of the base of Pike's Peak. A road will be laid out in the spring from the South Platte to the sources of the Smoky Hill fork. Yours truly, G. N. WOODWARD.

[B. HIATT] [209]

Auraria City, November 19, 1858

We have had a good time of it. The weather has been favorable and our health good. I have one share in Auraria City – a new place started at the mouth of Cherry creek, which is improving very rapidly. We have been here four days. There are seventy-five houses already put up, and three hundred more under construction. . .

I was forty-two days making the trip, and the roads are good. If you had fifty cows here you could sell them at big prices. Mexican flour is worth from $12 to $15 per hundred; bacon 25c per pound; common whisky $8 per gallon, retailing at 25 cents a drink.

The gold is here sure, and I mean to have my share of it. If you start with the mill bring five yoke of oxen to the wagon, and take some feed with you; start in March or April. I will write when ever I can send a letter. Direct care Wm. Smith, Leavenworth City, as he has started the express line to Auraria. . . B. HIATT.

-----

[209] *Missouri Republican,* January 6, 1859, copied from the *Leavenworth Times.* The same letter appeared in the *Kansas Tribune,* January 6, 1859.

[LETTER OF A.G.B.] [210]

Auraria City, K.T., mouth Cherry creek, Nov. 19, '58

Mr. M—: I arrived here on the 11th instant, having made the trip in forty-two days. We had a very pleasant trip, with the exception of six or seven days, on this end of the road. We traveled through the snow several inches deep; we have located in this town, and have four lots each, in the best part of the city, sixty feet by one hundred. I think this will make us pretty well off; and in addition to this we have taken up a claim of 160 acres of land each on Cherry creek – all good land. I have not prospected much, but am perfectly satisfied that there is as much gold in these mountains as there is in California. Men are making from eight to ten dollars a day with a pan. This is as good digging as I want. . .

As soon as the weather will permit, we intend to commence mining, and I am confident that I can make money.

At this time there are about 500 men in this vicinity. All are well pleased with the prospect. Yours truly, A.G.B.

[THOMAS L. GOLDEN] [211]

I send you a specimen of our gold, which I dug myself. I have discovered new mines, twelve miles from Cherry creek, on a creek called Clear creek. The creek empties into South Platte, ten miles below where Cherry creek empties into the Platte.

There are at this time fifty-three men at work at

---

[210] Kansas City *Journal of Commerce,* January 4, 1859.

[211] This undated letter must have been written about November 20, 1858. It was published in the *Missouri Republican,* January 6, 1859. Golden was one of the founders of the towns of Golden and Golden Gate.

these mines,[212] who average from $4 to $10 per day. Several old miners are at work in the mines. They say they are satisfied these mines are as good as any of the mines in California. We have prospected ten miles square. It will all pay wages by bringing water to it. We have organized a company of one hundred and begun a ditch, which we will complete by spring. It will afford plenty of water to work all the dry diggings. Yours, THOS. L. GOLDEN.

[GEORGE SEVILL] [213]

Auraria City, Cherry creek, Nov. 20, 1858

We arrived safely here, making the trip in 42 days. We have a house built for winter quarters, and have each got a share in the city.

We are unable to mine any this winter as the weather has begun to turn cold, but the gold is here and no mistake. They are now digging along the South Platte. . .

I wouldn't take $1000 for my lots in this town. I would like to send you a specimen of the gold but we have not been able to mine as yet. Farewell, GEORGE SEVILL.

[BUESCHE] [214]

South Platte, near Pike's Peak, Nov. 20, '58

Dear Zeitz: — After a very pleasant journey we have

212 Here, about two miles east of present Golden, Arapahoe City was begun late in 1858.

213 *Leavenworth Times*, January 1, 1859.

214 Published in the *Western Argus* and copied in the *Missouri Republican* of December 29, and in the Kansas City *Journal of Commerce* of December 31, 1858. The *Missouri Republican*, in introducing the letter, says that Buesche was long a resident of Chicago and is well acquainted with the Germans there.

reached the new El Dorado, in fine spirits and health. Our first business was to build a house, which kept us busy for several days – it being in size eighteen by twenty feet. The logs had to be brought from over the river, which retarded our work very much. As soon as we had our house up, we set about making the acquaintance of Americans in the neighborhood. We fell in with a company among whom resides the only surveyor of this far distant country. We took a trip up the river, where we found a beautiful situation for a town. The place is at the mouth of Plumb creek on the South Platte, in a most magnificent valley, at the entrance of the mountains. Our company surveyed a piece of land of one square mile, and we have already laid the foundations for several houses. As we expect an officer here soon, we think that he will register this land for us.[215]

All those who call on you for information in regard to the best route, should be advised to go via Bent's Fort, on the Santa Fe road, to Pike's Peak, which is 150 miles shorter than other route. As to the fact of there being gold here, and in some places in very great quantity too, there is not the least doubt. Until now we have done nothing but build our house and prospect. We have just commenced making sluices.

This morning we made four claims of 50 by 100 feet, which is the usual size of mining claims here, where we can bring our long toms in use and where we took thirty cents worth of gold out of two pans of earth. Yesterday one of our party, who never washed gold before, made $2.50 out of ten pans of earth.

There are from three to four hundred people here. . . Yours truly, BUESCHE.

---

[215] This was to have been Pike's Peak City.

[OSCAR B. TOTTEN] [216]

Auraria City, K.T., mouth of Cherry creek,
Nov. 20, 1858.

Friend Lusk: According to promise I avail myself of the opportunity of writing you a few hurried lines by the express, which leaves here in the morning for the states, to give you some idea of the new Eldorado, having arrived here on the 11th. . .

That this will prove a second California, I have no doubt; and I think we will have a population of some 50,000 persons by next fall. All that are now here are satisfied with the prospects and the country. The farming land is good, rich in soil, . . . Yours truly, OSCAR B. TOTTEN.

P.S. – The St. Louis boys are all here, and well, and perfectly satisfied. The following are the names of our party: Wm. H. Parkinson, Thoe. Parkinson, A. B. Reid, John Scudder, John Harris, S. B. Bassett, C. H. Noble, Albert G. Baber and your humble servant. O.B.T.

[JOHN SCUDDER] [217]

Auraria, Nov. 24, 1858

The St. Louis and Pike's Peak Mining Company,

---

[216] Published in the Kansas City *Journal of Commerce*, January 18, 1859, and copied from the *Jefferson Inquirer*. Totten became a Colorado pioneer. He was active in political matters in 1859, being secretary of the convention that formed Jefferson territory. He was listed as a lumber dealer at Denver in the fall of 1859. His portrait appears in *Representative Men of Colorado*, 181.

[217] Published in the Kansas City *Journal of Commerce* of January 20, 1859; copied from the *St. Louis Republican*. Scudder was elected treasurer of the Auraria Town Company in April, 1859. On April 16, 1859, he had a quarrel with P. T. Bassett, whom he shot and mortally wounded. Scudder left Denver and went to Salt Lake City, where he engaged with W. H. Russell and helped outfit the pony express. He returned to Denver in 1860, where he was tried for the killing of Bassett and was acquitted. He was living in Denver in 1903.—Smiley, *History of Denver*, 250.

which left St. Louis in September in search of gold,
reached this place on the 15th of November, after
being fifty days on the route. Our company consisted
of six men: J. T. Parkinson, James Reed, John Harris,
Chas. Dallah, P. T. Bassett and myself. We found on
arrival here, about 100 men, who had come before us,
and commenced to build a city. The arrivals from that
time have been about 15 or 20 each day. There are now
erected about 35 good log houses, and if the snow does
not come too soon, in one month from this date the city
will number at least 150 cabins, with 600 inhabitants.
. . Truly yours, JOHN SCUDDER.

———

[J. S. HARDIN] [218]

Cherry creek, November 24, 1858

. . . There is a man running an express from here
to Leavenworth and Kansas City. He charges one dollar for every letter he carries. There is also an express
to Ft. Laramie, which charges 50 cents for the first
letter and 25 cents for each additional. . .

There are some four or five hundred men here. There
is a town at the mouth of Cherry creek, with some fifty
or sixty inhabitants – some three or four women
amongst them; one a Mexican, one or two Mormons,[219]
and one Indian. The Indian is the wife of an American
white man. The Mormon ladies are from Utah, and I
believe the Mexican is from New Mexico. There are
perhaps twenty-five Mexican men amongst the mess. . .

(Men have held an election for a separate territory.)

Since then, within the past few days, two men arrived
here, claiming to have been sent from East Kansas, to
act as sheriff and judge; with orders to lay off a county,
locate a county seat, and, I suppose fill the offices to

218 *Nebraska Advertiser,* January 20, 1859.
219 Mrs. Samuel Rooker and daughter.

the best advantage; but by some means, the men of the
town below have told them that they would not like
to live under the Kansas government, and I hear threatened to tar them.

[SAMUEL S. CURTIS] [220]
Winter quarters, 2½ miles below Cherry creek.
Nov. 24, 1858.

Mr. Editor: We are at length comfortably housed
for the winter, and have been for the last few days engaged in looking over the country in this vicinity. . .

Fort St. Vrain is about 50 miles below here. It is a
well built adobe fort, about 100 feet square, with but-
tresses on the south east and north west corners. About
seven miles south of it there is another fort, and one
about six miles this side of that.[221] They are all about
the same size, and appear to have been abandoned at
about the same time. Sander's ranche was the next
object of interest, as there we first reached the Cherry
creek settlements. The bottom is quite wide here, forming a large island which is covered with good grass and
young cottonwoods.

Sanders has been a mountaineer since '44 and has
herded cattle at this place for five winters. There are
several persons connected with him in the business,
among whom I found a nephew of our friend Col.
Cochran. They start an express to Fort Laramie on
the first of December, and by it this letter goes to the
post office. All letters to this vicinity should be directed
to Fort Laramie, in care of Sanders and Company's

220 *Council Bluffs Nonpareil*, January 22, 1859. See above for an earlier
letter of Curtis.

221 This is welcome data on the bastions of Fort St. Vrain. The next fort
to the south would be Fort Vasquez (1½ miles south of Platteville) and the
second, Fort Lupton (near the town of that name). For data on these forts,
see the preceding volume in this series, page 143.

Express to Cherry creek. The express will run every month, and Sanders is the man to put it through.[222]

Towns are springing up all around us. Three are already started, and three more are projected, Montano, Auraria, and Denver City are already in operation. Denver is the county seat. We have a full set of county officers, appointed by the governor of Kansas.

There has been but little done, as yet in the mines. . . I have been to the mountains, and in the first pan full in which I obtained the "color" I got about 3½ cents worth of gold. Further examination showed still better results, and now there is quite a number of claims taken there. The bed rock is about fifteen feet below the surface of the ground, and about the same distance above the creek. The surface pays about one cent to the pan; next to the bed rock, about five, and the bed of the stream, in some places, as high as thirty. The place I speak of, is where Ralston's creek leaves the mountains.

Yesterday, the 26th, a train came in from Sioux City, Smithland, and that vicinity. . . Yours truly, SAM S. CURTIS.

[L. J. WINCHESTER] [223]

Golden City,[224] Nov. 25, 1858.

Dear Major: We arrived here safely after a pleasant trip of 46 days from Kansas City, being the first party out this fall. On our arrival we found some 16 or 20 persons here only. We at once proceeded to find out if there was any gold here, and to our great joy and sat-

---

222 For further data on Sander's express, see L. R. Hafen, *The Overland Mail*, 145-146.

223 *Omaha Times*, January 20, 1859. Winchester was a stockholder in the Denver City Town Company. He attended the first Masonic meeting in the region on December 10, 1858.

224 Apparently this was Denver. The name "Golden City," was applied to the town for a day or so. See Larimer, *Reminiscences*, 89. The present city of Golden was founded in 1859.

isfaction, found we could dig nowhere without finding more or less of it. The mines which are now being worked, are yielding from $3 to $5 per day to the man, worked with cradles or rockers. These are not the richest mines by any means, but are the only ones that can be worked this winter.

We are located on the south fork of the Platte river, in a lovely country. . . Yours truly, L. J. WINCHESTER.

[G. N. HILL] [225]

Aurora, mouth of Cherry creek, Colona territory, Nov. 28, 1858

This town is six weeks old, and contains 107 houses. Across Cherry creek is Denver City, laid out by the county officers sent here by Gov. Denver. This town has something like a dozen houses, all told. Five miles above here is another town called Montana, laid out by the Lawrence company, also Arrapahoe City;[226] Still higher up is Santa Fe.[227] These are all the towns I have heard of since I have been here. . .

Now for a word on political matters. Gov. Denver appointed and sent out here a board of county officers to administer justice to us.

We thought we were out of the pale of civilization, and would be allowed to regulate our domestic affairs to our own liking, but it seems we could not be allowed that privilege.

The citizens of South Platte held a meeting and requested the officers to resign, and let them elect their own rulers. They took time to consider the matter, and

---

225 Kansas City *Journal of Commerce,* January 15, 1859.

226 This projected town was two and one-half miles south of Montana, according to A. O. McGrew's letter of December 29, 1858, published below. The Arapahoe City that was founded in the fall of 1858, and which survived for several years, was located on Clear creek, about two miles east of Golden.

227 This was apparently a paper town. Its location has not been determined.

have not yet given their decision. If they do not resign, look out for squally times in the valley of the South Platte.

[JOHN GRAVES] [228]

Auraria, K.T., Nov. 28, 1858

. . . I have been here now a little over a month, but have done little but prepare for winter. This is now done, and we are now ready to prospect more thoroughly. We have done a little of it already, and I am pretty well satisfied that there is gold here in sufficient quantities to pay for mining; . . .

The prices current of provisions here are as follows: flour, per pound, 30 cents; coffee, 35 cts; sugar, 35; bacon, 40; salt, 20; tobacco, $1.50 to $2.00; saleratus, $2.00; rice, 40; beans, 25; whiskey, per gallon, $8; molasses, $4.00. Yours truly, JOHN GRAVES.

[JOHN J. SHANLEY] [229]

Auraria, C.T., Nov. 30, 1858.

. . . I arrived here about the 29th of October, after a pleasant trip – the weather, which is, after all, the main blessing, having been beautiful.

The prospects here are flattering. The various stories we heard before leaving, detailing the richness of the mines, are, I think, as near correct as could be. . .

We are located at Auraria, and each one of us owns an original share in the "afore-mentioned." When we first arrived here, (four weeks ago,) there were but two houses finished, and two under way. There are now here about two hundred and fifty men, fifty houses built, and half as many more will be finished within the next sixty days, and about as many foundations laid

228 *Council Bluffs Nonpareil,* January 22, 1859.
229 *Omaha Times,* January 20, 1859.

which will be built upon in the spring. As there are finished and unfinished houses, so you see we are not by any means "country Jakes," but live in a city. Yours truly, JOHN J. SHANLEY.

[HENRY ALLEN] [230]

Aurori [Auraria], N.T., Dec. 1, 1858.

Dear Wife and Daughters: I am now seated in my own log cabin, and if I only had you all here I would be truly happy. I hope by next summer to have you here. There is but two white women in this country that I know of. We have had a large deputation of Arapaho Indians on whose lands we are. They are very friendly, and appear to be pleased that we are here.

I have been about thirty miles south of here on Cherry creek, prospecting for gold, and panned out forty-five cents worth in one pan full of dirt. It is very fine. I think with the right kind of tools a person can make from five to twenty dollars per day.

Wagstaff's boys are about thirty miles below here; but I have not seen them. I understood they brought a letter for me which got lost on the way, and I am sorry that it got lost as I have not heard a word from the Bluffs since I left. I have just completed a bedstead

---

230 *Council Bluffs Bugle,* January 26, 1859. In introducing the letter this newspaper says that Allen had been a legislator in Iowa, was an alderman and the postmaster of Council Bluffs before his journey to the mines. He became postmaster of Denver, was prominent in the provisional government of Jefferson territory and in Masonic activities in Denver. He brought his family out in 1859. The marriage of his daughter Lydia to John B. Atkins on October 16, 1859, was the first such ceremony performed in Denver. See the *Rocky Mountain News,* October 20, 1859. In the 1860's Henry Allen went to Idaho and then to Montana. In these territories, as in Colorado, he was prominent in Masonry. He died in Los Angeles, California, February 18, 1871. See George B. Clark, *Our Masonic Heritage* (Denver, 1936), 59.

A letter written by C. H. Gibson and dated at Cherry creek, December 1, 1858, was published in the *Chicago Press and Tribune* of February 3, 1859. He says he arrived November 14. He says miners are making from $3 to $8 per day.

for Jim and me, and have filled the bed-tick with prairie feathers (grass). It is now snowing hard, and it looks like we were a going to have a deep one. Jim has just wrote to Willey.

In this letter I send you a little gold dust. It is worth about seventy-five cents. This I got out in two pans of dirt. Show it to Dr. Craig; he can tell whether it is as good as California gold.

I know where there is diggings that the gold is larger than this. I have seen some about the size of common white beans; but it is impossible to do anything like work in the winter; . . .

While I am writing to you half a dozen Arapahoes are sitting beside me talking their language and calling me chief. I have been elected director of the town company and president of the claim company,[231] and the Indians seeing whites coming to me on business, think I certainly must be a chief and have dubbed me chief accordingly. One thing I want you to be sure and tell all our friends that come out in the spring, that old Californians have no advantage over the rest of us, from the fact that the diggings here beats them out — they are not confined to any particular location; but are scattered all over the country, for a distance of over two hundred miles. The whole country appears to be impregnated with gold.

We have organized claim laws for farming and mining, which I think are very good and liberal. We have preaching every other Sunday by a Methodist brother by the name of Fisher — a very clever and good man; tho' "a fisher for souls," he is here like the rest of us fishing for the "filthy lucre." Those who come out should start early in the spring.

---

[231] The records of this important claim club have not come to light. They may have been lost in the Cherry creek flood of 1864, when the Denver city hall was washed away.

There is about eight hundred persons in this country.

If the girls could only be here to see the Rocky mountains in the morning or evening, when the sun is shining – they would see one of the most beautiful sights they ever beheld. It is about 8 miles to the mountains. The first tier is covered with pine timber, and looks as black as can be. The second tier is covered with snow the whole year round; and looking at both, with the sun shining upon the peaks of snow, the craggy rocks, and upon the pines below, forms the grandest sight I ever saw.

It costs fifty cents to get a letter from here to Laramie, and the same to get one back from there. It used to cost a dollar.

We are going to have a train come in in July, and bring out our families. If any one comes out soon, write me a letter and send it by them – if not send your letters by mail to Fort Laramie.

I remain yours affectionately, HENRY ALLEN.

[NOEL LAJEUNESSE] [232]

Denver City, (Cherry creek)

Dear Father: As the express train leaves here tomorrow morning for Laramie, I take the opportunity of sending you these few lines. Times here are brisk, there is three towns laid off here, . . . Horses are very scarce here. I wish you would tell my brother Mich. to come over with Mr. Papan and bring me over six good horses. Government officers appointed by Gov. Denver of Kansas, commissioners, supervisors, judge of probate, and sheriff arrived here. Yesterday there

232 This undated letter was probably written on December 2. Later in the letter the writer says that Sheriff Wynkoop would leave the next day. The sheriff set out December 3 and reached Omaha January 5 (See *Southwest Historical Series,* IX, 210). Lajeunesse's letter was printed in the *Rulo Guide* and reprinted in the *Nebraska Advertiser* of January 27, 1859.

was a large meeting of the people who proclaimed them legally commissioned officers; bye the bye, the present sheriff is going to the states to-morrow, having resigned his office; Jack Jones [233] has been appointed to take his place, and will take the oath of office this morning. Through the influence of Jones I have become an equal stockholder in two towns, which will be worth ten thousand to me next summer. If my brother comes over as soon as possible, he may stand a good chance of making a good thing out here yet.

Come out here as soon as possible in the spring, and bring out fancy groceries, fine liquors, boots and shoes, (large size) and you will find a ready sale. . . NOEL LAJEUNESSE.

[J. B. WISENALL][234]
Winter quarters, 3 miles north of Cherry creek, Dec. 3, 1858.

. . . We reached this place on the 30th of October, just in time, for we had a snow storm the same night. . . Six teams were here when we came; there were seven of ours; eight more came the same day, and eleven three days after – so we have quite a village, of twenty-five houses and one hundred men. Clancy, of De Soto, came with the last party, and is camped just on the opposite side of the river, which is about 100 feet wide. . .

---

[233] Also known as William McGaa, who was married to a half-blood Sioux woman. His son, William Denver McGaa, was the first child born in Denver. See J. C. Smiley, *History of Denver,* 244-245, for his picture and story.

[234] *Omaha Times,* January 13, 1859. In the same newspaper is a letter from D. F. Richards, dated at Auraria, December 4. It contains no important information. A. Vinnance's letter of December 4 appears in the *Omaha Times* of January 20, 1859. He gives a flattering account of the mines, saying that the Mexican miners, with wooden bowls, are making from $5 to $15 per day.

I am sending this letter by a Mr. Steinberger, who is going in. Teams leave nearly every week for the states, with the intention of bringing stores, mills, etc., in the spring. . .

I remain, yours respectfully, J. B. WISENALL

[W. W. SPALDING] [235]

Pike's Peak City, Dec. 7, 1858.

Editor Journal of Commerce: Dear Sir: As there are two men from our city starting home to Indiana, and will be in or near Kansas City, and as I have a chance to send a letter to you I will give you all about the sayings and doings of this country.

Pike's Peak City is a new laid out town fourteen miles from the south fork of Cherry creek, and a fine locality for a prosperous town, it numbers about 38 cabins and eighty-two men and one woman. We have had a great deal of snow for this time of year but not much very cold weather.

There will be a brisk time in this part of the country next spring as all the miners that are here at this time, are hunting out the richest claims for the spring working. Some are making sluice boxes, some long toms and cradles. I saw Jas. Winchester and J. J. Price yesterday. They have been prospecting for spring diggings for three days, and they found what they thought were good diggings, if the snow and water were gone. They had made $41.20 in three days prospecting with a pan. The miners that work steady all the time make from $1.50 to $8.00 per day, with pans, and $4 to $12 with rockers and very rough ones at that. . . W. W. SPALDING.

---

[235] Kansas City *Journal of Commerce,* January 14, 1859.

[THOMAS WARREN][236]

Montana, Dec. 11, 1858.

Thomas West, Esq., Kansas City. Dear Sir – Since my last letter to you, I have arrived here in the heart of the gold region, and am now able to give you my own impressions. I am snugly cabined for the winter, and, with the exception of a frozen foot, am as comfortable as heart could wish. My anticipations in regard to this country have been more than realized, and although the winter has been of such a nature to preclude the possibility of mining to any great extent, still I have seen enough to justify me in believing that gold exists in great abundance, and that the opening of the spring will disclose its rich treasures to us. . .

Two flourishing towns are laid off five miles below this, at the mouth of Cherry creek, one called Auraria, signifying Golden Light, and the other Denver City. Two and a half miles above this place, is another town known as Nonpareil City,[237] so that you see we are well supplied with towns.

I suppose that I do not exaggerate when I say that at least 500 buildings are now in progress of erection in Denver City and Auraria, in anticipation of a general rush in the spring.

A saw mill is much needed here, and would prove a rich speculation to the proprietor. Inferior lumber, ripped out with a whip saw is worth $25 per hundred feet at present.

A very large ice house is being built here, and our folks will have something to cool them off during the heat of summer. . . Yours, THOS. WARREN

---

[236] *Ibid.,* January 29, 1859. J. C. Smiley, *History of Denver,* 244, says that Thomas Warren was a Kentuckian and the first operator of a ferry across the South Platte at Denver.

[237] This was another paper city.

[From a letter [238] written the following day, we quote:]

Mr. Burton: Dear Sir: Prices here are exorbitant, and will be higher in the spring. Good clothing, provisions, medicines, building hardware, sheet iron, books, stationery, and such articles will sell readily. All kinds of mechanical trades will do an excellent business here. At present we have but one watch maker, one gun smith, three blacksmiths, and some few carpenters among us, and they are steadily employed, although they have but few tools with them. THOMAS WARREN.

[A.F.B.] [239]

Auraria, Dec. 14, 1858.

Messrs. Editors: – As it is snowing and very cold, I thought it would be a good time to write and give you the news from this Eldorado. Since, I have been here, a little over a month, I have prospected considerable. The first trip I went south about thirty miles from here, and panned out in several places from two to twenty cents to the pan, or three spades full of dirt, and there was not one place on the creek where I did not get the color.

On Ralston creek, north-west from here, we done about the same, and on Platte river, near the mountains, we did a little better.

Capt. Cook's company, of St. Joseph, are now running a "tom" on Ralston creek and are making from five to six dollars per day. . . It is impossible for us to enter the ravines or canions on account of snow. Some have tried and have paid dearly for it, by being

---

[238] Kansas City *Journal of Commerce*, January 28, 1859. In introducing this letter, the newspaper speaks of Mr. Warren as a most exemplary young man. He lived on a farm near the city, in the region called "Goose Neck." The letter was copied in the *Lawrence Republican* of February 3, 1859.

[239] *Council Bluffs Bugle*, February 23, 1859.

badly frozen – some have had their feet cut off, others their toes. . . I suppose that you have seen H. J. Graham on his way to Washington, and as a matter of course he has given you all the political news. Since he left we have had some gentlemen sent amongst us as officers of Arapahoe county. One is of rather a strange name (Smith), district judge, Gen. Larimer, (I think you have heard of him before) treasurer, E. Wynecoop, sheriff, and three others that claim some office, but I have not found out what they are. At first the people refused to receive them; but out of respect for Gov. Denver, who appointed them, we did receive them; but there is still considerable dissatisfaction, and I fear that it will not work well. Men sent out among people like we are, to administer the laws, should be men of good character. I am sorry to say there is considerable drinking and humbugging going on; but it is confined to a certain class.

Blake and Williams, formerly of Crescent City, have a small stock of goods here and are selling at good profits. Whiskey that will kill at forty-five rods, eight dollars per gallon; that will kill at twenty rods, sixteen dollars per gallon. Flour, (Taos, 140 lbs) twenty dollars, and that from the states, (100 lbs) twenty-five dollars per sack. Bacon, 40 cts per pound. A.F.B.

[JOHN CUSSONS] [240]

Waw-zazh-ee (near Cherry creek,) Dec. 15, 1858.
. . . Here is the burden of my communication: There is gold, and mining will pay well to those who are able and willing to work, and to such, I say, by all means, come. But to lawyers, paper-city brokers, doctors, delicate clerks, politicians, and bar-room loafers,

---

[240] *Missouri Democrat*, February 1, 1859, copied from the *Herald of Freedom*.

I speak the sentiments of the people when I say, we have no use for you.

Three cities were established here in one week: San Francisco, Sacremento and Auraria; but it pains me to record the demise of the two former, they having "bust up" from a lack of that, without which no city can prosper – houses. Since the last fall of snow the "proprietors" are somewhat dubious of the whereabouts of their cities, but entertain strong hopes of finding them as soon as snow thaws.

The city of Auraria is in a flourishing condition, numbering more than one hundred neat log cabins. . .

The lands on the margins of the streams are admirably adapted to agricultural pursuits – particularly the cultivation of roots and vegetables; but the heavy hail storms, so common to this country, will prove very injurious to grain. The season is almost too short to mature corn, consequently early kinds should be selected for seed. The country abounds in game, such as black-tailed deer, antelope, hare, rabbit, turkey, pheasant, partridge, prairie fowl, etc. Immense herds of buffalo range here during severe winters. . .

Cattle are the beasts of draft for the prairie, they will draw more, travel faster, and live on less than mules. They should not be less than five years old, and well broken. JOHN CUSSONS.

[HENRY ALLEN] [241]
Aurori [Auraria], N.T., Dec. 17, 1858.
Dear Perin: – After a long silence, I have concluded that I have learned enough of this country to give you

[241] *Council Bluffs Bugle,* February 9, 1859. A letter from David Kellogg, dated at Montana, December 15, was published in the *Waukegan* (Illinois) *Gazette* and copied in the Kansas City *Journal of Commerce,* February 23, 1859. He tells of an old man who told him of finding gold ten years before. See the preceding volume in this series, page 33.

some information about it. I have been over considerable of the country and know that a person can do well
here. . .

First comes Aurori, about 60 miles N.W. of Pike's
Peak, and nine miles from the Rocky mountains, and at
the mouth of Cherry creek. It contains sixty houses,
and seventy-two more going up. On the east side of
Cherry creek stands Denver City – five houses up, and
five or six going up. Seven miles above is the town of
Mountana – seven houses. Thirty three miles up Cherry
creek is the town of Russelville, with five or six houses.
You will see we are well off for towns. I think some
of our boys are agoing it too strong on corner lots, they
may make something; but I had rather risk the gold
diggings. Although I am interested in all the towns,
I don't count much on them. There is a company about
eight miles from here that have commenced mining,
and the mines are paying well – Sam Curtis is in it.[242]
The Russelville diggings are good. I shall go up there
next week to work the mines while the weather permits.

Perin, I know you could do well here. James Wagstaff left yesterday for the mountains prospecting. There
is over nine hundred white men, and two white women
here, with a right smart sprinkling of Mexicans and
Indians, and we all live together friendly. When I get
to work at Russelville, I will give you the gold news.

Our home is now hung full of meat. Three of us
went out hunting, and was gone three days. We got
eight white-tailed deer, and four mountain sheep, and
left twenty-four deer that they could not haul in. In
fact, the mountains are alive with deer, antelope, and
mountain sheep. . .

James is now on Ralston creek prospecting with Judge
Townsley of Sioux City. Foster has gone to Cash la

---

[242] This was Arapahoe City, on Clear creek.

Poudre, where we have heard of shot gold being found. We are all well, and all satisfied we are going to make a fortune. So come if you can. I must write to A. D. Robinson tonight. I remain yours truly, HENRY ALLEN.

[OSCAR TOTTEN] [243]
Auraria City, K.T., mouth of Spring creek,
December 17, 1858.

Friend PARKINSON: Having an opportunity of sending letters tomorrow, I thought I would write you, and let you know how matters are going on in the good city of Auraria, near Denver City. I would say that the place is improving fast, houses are going up on every side, and there is a general activity among the people. As yet we have had but one or two cold spells. The weather at the time of writing this is mild and pleasant. We begin to present quite a city, there being now some seventy-five to eighty houses erected, and many more in progress. Every one seems cheerful, and contented with regard to the country and the prospects before them.

Denver City is going ahead, and will prove to be quite a place, before winter bids us adieu – that it is a most beautiful town site, you well know. It is reported here, that coarse gold has been found on the head of Long's creek [present Clear creek], and it has been found on Dry creek, which is a certain fact. That the whole region of country is auriferous, is now beyond all doubt, or cavil – and I predict that this at no distant day will be a rich and populous state – adding another bright star, to our glorious union. William & Mc-Fading leave day after to-morrow for the Arkansas, to

243 *Missouri Republican,* February 2, 1859. The letter is addressed to his friend Parkinson, who came out to the mines with Totten and had returned to the states.

secure its water privilege as also to lay out a town.

That Auraria is a moral place, and an exception to all new countries, especially gold countries, you will be bound to admit, as we have preaching every Sunday morning, and a temperance society started. Among the members are John Smith, William and your humble servant. I think it is a great credit, and it speaks volumes for the country and the city of Auraria. When you return we will enroll you among its members. . .

Dr. Russell intends going north to secure the water privilege of two streams which empty in the Platte, upon which he got good prospects.

The difficulties which hung over Denver City have all been settled; everything is working right, and all goes on well. Give our respects to enquiring friends.

Well, I will close, wishing you a pleasant visit to your friends – a merry Christmas and a Happy New Year, I remain, your friend, OSCAR.

P.S. – I came very near forgetting to tell you of a hunt which William McFading, Baker Noble and three others had last week, and not a good week for hunting either. They killed fifty-eight black tail deer, four mountain sheep and two catamounts, besides some few mountain grouse. It took one four-mule team, and one ox team of three yoke of cattle, to haul the game in. Put that down in your hunting book. It would rather surprise some of the old hunters about St. Louis to tell them that they were all killed in four days! But such is the case. Don't look like starving, does it? So all's well that ends well. O.

[D. C. COLLIER] [244]

Denver City, K.T. Dec. 18, 1858.

On arriving here we were greatly, but very agreeably

---

[244] *Lawrence Republican,* February 10, 1859, copying from the *Wyandott*

disappointed in finding a town of not less than three hundred inhabitants – the growth of less than one month. The bright, golden prospects, which are presented by the whole country round about, have occasioned the laying out of this town, which is built after the style of the early days of Sacramento and San Francisco. There are now here about one hundred houses completed, and as many under way. Our houses are built of logs, often finely hewed, with ground floors and flat roofs, made of poles covered with grass, or rushes and earth, with sod, stone and stick chimneys. These make us comfortable dwellings for the winter, but will be succeeded by better ones, as soon as we can obtain mills to supply us, and as soon as brick can be made. A good quality of pine lumber, cut out with pit-saws, is now selling here for twenty cents per foot.

The best diggings yet found are said to be those on the Cache le Peadre [Poudre], Vascarris [Vasquez] fork, and those near Table mountain, together with those on the Bijou Solon [Bayou Salad, South Park?]. But little has been done in any of those places yet.

The most of the digging is at present being done within a few miles of this place, on the Platte and on Dry creek, where there are thousands of acres that pay, with pans, from two to ten dollars per day. Larger accounts than this are often given, and I, myself, know of one reliable case where $25 was taken out in one day. The universal opinion is, however, that the best

Gazette. Collier was born in New York state, October 13, 1832. He graduated from Oberlin College in 1857. After coming to Kansas he joined the gold rush, setting out from Kansas City, October 7, 1858. He became a lawyer and real estate dealer in Denver. In 1862 he moved to Central City, where he practiced law and edited the *Central City Register.* His biographical sketch in the *History of Clear Creek and Boulder Valleys, Colorado* (Baskin, 1880), 444, says he reached Denver on December 5; the *Roll of Members of the Society of Colorado Pioneers* (published in 1883) lists Collier as arriving on November 1, 1858.

diggings, – the source of the gold – is yet to be found.

As yet, cradles, toms and sluices have not been much used. D. C. COLLIER

Since writing the above, new discoveries have been made, where from twenty-five to thirty dollars can be made per day with pans. Gold bearing quartz has also been discovered, and is said to be abundant. The new diggings are in the mountains, about twenty miles distant. C.

[H. M. UMPHREY] [245]

Auraria, K.T., Dec. 29, 1858

. . . We reached Cherry creek valley on the 6th of December, after a tedious journey of two months. Two weeks of this time we were compelled to lie over, and we were further thrown out of our way and detained by want of a proper knowledge of the route.

Those who return from here to the states will mostly take the northern route down the Platte river; but I think there will be a road opened up through the Smoky Hill valley in the spring, which will be the most direct and best route to the mines.

This town is situated in the forks of Platte river and Cherry creek, and contains seventy or eighty log cabins now occupied, and many more are in progress of erection. It has one store. A large train, loaded with goods and provisions, has just come in from New Mexico. Previously, flour had been selling at $25 per 100 lbs; it can now be had for $15 to $20. Whiskey sells for $8 per gallon, and other things in proportion. It is estimated that there are seven hundred men on the Platte river and the creeks in this neighborhood. . .

As to the gold – I believe it is here over a large

[245] Published in the *Alton Courier* and copied in the *Herald of Freedom* of April 2, 1859. Umphrey was from Alton, Illinois. The letter was written to his brother, Madison Umphrey of Alton.

extent of country in length and breadth; but has not yet been found in sufficient quantities to pay well for the working. . .

There are a great many men here laying out towns, and making a great "blow" about the gold that is to be found here, for the sake of encouraging emigration, and speculating on their town property; but I warn you to beware of the glowing and enticing accounts sent out from such sources. . . H. M. UMPHREY

[CHRISTMAS CELEBRATION, 1858. A. O. McGREW][246]
Denver City, Dec. 29, 1858.

Editor of the Omaha Times: – A brief retrospect of occurrances since my sojourn in this portion of the country; a short account of things as they are; a description of our Christmas festival; Masonic dinner, and a few of the leading events of the day, including the new gold discoveries, alum mines and other deposits, and the prospects of the new towns, may not be the least interesting news matter set before your many readers, and the departure of Sander's & Co's express for Fort Laramie offers a favorable opportunity for sending news to the states, or, as some of the boys facetiously term it, "to America," I shall jot down such information to those who contemplate coming here in the spring. . .

Our drives were as follows: We left the river on the 27th of September last, and drove to Papillion creek the first day; second day to Elkhorn, 3rd day to Rawhide creek; 4th day five miles west of Fremont;

[246] *Omaha Times,* February 17, 1858; republished in the *Colorado Magazine,* XIV, 15-25. McGrew was from Pennsylvania and was a correspondent for certain eastern papers. He was known among Colorado pioneers as "the wheelbarrow man," because of his attempt to reach the mines with a wheelbarrow. After having pushed it about halfway across the plains, he joined a wagon company, and his vehicle was thereafter used for gathering buffalo chips for the company fire.

5th day, 63 miles from the place of starting, to a place called Buchanan; 6th at Columbus; 7th day at Loupe fork; 8th day on Prairie creek; 9th day, 20 miles west of Prairie creek; 10th day, 21 miles further west; 11th day, 20 miles further west; 12th day, camped on Wood river; 13th day, camped opposite Fort Kearny; 14th day, 17 miles west of Kearny; 15th day, Plum creek; 16th day, 17 miles further west; 17th day, laid over; 18th day, drove 18 miles; 19th day, camped on Cotton-wood; 20th day, at Cottonwood springs; 21st day, drove ten miles; 22nd day, drove 17 miles; 23rd day, camped at Big spring; 24th day, drove to the crossing of the South Platte; 25th day, drove 18 miles; 26th day, drove 18 miles; 27th day, drove 18 miles; 28th day, drove 20 miles; 29th day, drove 18 miles; 30th day, camped at noon on the Bijou, and drove that night to Kiowa creek; 31st day, drove ten miles; 32nd day, drove 16 miles; 33rd day, drove 1½ miles; 34th day, drove 8 miles and camped; 35th day, drove 16 miles; 36th day, we made 12 miles; and camped within a mile and a half of our present location. The next day we moved up and commenced building our present spacious and commodious residences, which being completed, were occupied in a much shorter space of time than it takes a fashionable establishment in New York to set it-self to rights. . .

We are now at the mouth of Cherry creek, the goal, to which so many are looking with such intense anxiety in the spring. Here we find already, two flourishing towns – Auraria, on the west side, and Denver City, on the east side. The latter place has been selected as the county seat, and is even spoken of as the capital, in case a new territory is stricken off. Five miles above Auraria, on the Platte, is the mining town of Montaina; two and a half miles above is Arrappahoe City; and a mile and a half farther up is the Plumb creek settlement. Three miles below Denver City is Curtis' ranche, and on the opposite of the river is Spooner's ranche, our present place of residence. Chat. D'Aubrey's trading post is three miles below; nine miles farther down is Sander's ranche; five miles further is Fort Lancaster [Lupton]; two miles further is Fort William, formerly Lupton's ranche; ten miles further down is St. Vrain's Fort; and five miles further down is Cache la Poudrie, where the shot gold is found. The latest and richest gold diggings found, are at the base of what is known as Table mountain, where, already, a town, known as Mountainvale [Arapahoe City], has been laid out and several houses erected while others are in progress of erection. At the head of Cherry creek, the town of Russelville has been laid out, and a large number have gone to commence building operations. This comprises the list of towns and settlements.

In regard to mining, the operations in that line have not been carried on to any great extent this winter, on account of the streams being either frozen or dry. . .

Talking of Christmas, Mr. Editor, have you poor frozen victims of the states, the remotest idea of what a real old-fashioned Christmas celebration is, and would it hurt your feelings or spoil your appetite, if I were to rehearse some of the sayings and doings of that eventful day; and set before you our "bill of fare" on that occasion, not an imaginary one, either, but a faithful record of the luxuries under which the board groaned — enough to tempt the appetite of the most fastidious gourmand? And as the day waned, and the good cheer disappeared, wit and sentiment flowed as freely as the sparkling wine, from which they emanated.

In order to have the affair come off with as much eclat as possible, a meeting was called to make suitable arrangements. The meeting was called Dec. 21st, and I make the following extract from the minutes of the secretary:

At a meeting held at Camp Spooner for the purpose of making arrangements for celebrating Christmas day in a suitable manner, J. S. Lowry, Esq., was called to the chair, and Hon. Wm. Clancy appointed secretary. Mr. Lowry having in a brief speech set forth the object of the meeting.

Mr. E. Matthews [moved] that committees be appointed, whose business it should be to make the necessary arrangements.

Carried.

On motion of Mr. J. C. Latta, Messrs. Spooner, Perkins, and Long were appointed a committee of arrangements.

On motion of Mr. E. H. Warner, Messrs. Baker, M'Lachlin, and M'Grew were appointed a committee on resolutions.

On motion of Jno. Burssee, Messrs. Franklin, Way and Stevens were appointed a committee on toasts.

On motion of Mr. Perrin, Messrs. Forbes, Hobbs and Dumont were appointed a committee on invitation.

On motion of Mr. E. Hay, Messrs. Frary, Sullivan and Orum were appointed chief cooks.

Among the invited guests, were General Larimer, E. P. Stout, Esq'r, S. S. Curtis, and about fifty others. The day, as I said before, was as bright and beautiful as ever shone upon the face of animated nature; not a breeze whispered through the leafless branches of the trees, and everything seemed to be enjoying a general repose. The boys, with their invited guests, lolled lazily around on the logs, smoking their pipes or spinning innumerable yarns about their gold prospecting and hunting expeditions. The cooks, or rather, "culinary professors," were steaming over their different departments, now turning a choice saddle of venison, then looking to the pastry to see that it was done to a nicety. One was moulding the fresh rolls of butter into all kinds of fantastic shapes, another cracking the nuts, so as to be ready for mastication. Anon, a head would protrude from a cabin door, with a cry of "more wine for the pudding sauce," would again disappear within those precincts, which were at least sacred for the time, to the labors of the disciples of "Soger." But, I am forgetting the dinner, which was the great feature of the day. The guests having taken their seats at the board, Hon. William Clancy was invited to take the head of the table; on his right was General Larimer, and on his left, Dr. Steinberger, who acted as secretary. Dinner having been fully discussed, the chairman called the board to order, and while they are coming to order, I will tempt you with the bill of fare:

Platte River Gold Diggings Bill of Fare.
Christmas————1858.

Soups
Oyster soup. Ox tail.

Fish
Salmon trout, with oyster sauce.

Boiled.
Corned beef, buffalo tongue, mutton, pork, ham, beef tongue, elk tongue.

Roast.
Venison, a la mode; buffalo, smothered; antelope; beef; mutton; pork; grizzly bear, a la mode; elk; mountain sheep; mountain pig.

Game.
Mountain pheasants; mountain rabbits; turkeys; ducks; sage hen; prairie chickens; black mountain squirrel; prairie dog; snipe; mountain rats; white swans; quails; sand hill cranes.

Extras.
Potatoes baked; potatoes, boiled; rice; beans, baked and boiled; beets, squashes, fried; pumpkins, stewed.

Desert.
Mince pie; currant pie; apple pie; rice pie; peach pie; mountain cranberry pie; tapioca pudding; bread pudding; rice pudding.

Fruits.
Brazil nuts; almonds, hazel nuts; filberts; pecans; wild currants; raisins; prickly pear; dried mountain plum.

Wine List.

Hockheimer; madeira; champagne; golden sherry; cherry bounce; hock; Monongahela whiskey; claret; brandy; Scotch whiskey; Ja. rum; Bourbon whiskey; Taos lightning.

If you have done with the bill of fare I will trouble you for your attention while the meeting proceeds. Order being restored, Mr. Clancy read the following resolutions, which on motion were incorporated with the proceedings of the meeting:

Resolved, that we hereby tender our thanks to Capt. R. A. Spooner, for the able and efficient manner in which he conducted our train through from Nebraska to this point, and for his kindness in everything that contributed to the comfort and happiness of all the members of the company. Resolved, that we also tender our thanks to our secretary, Dr. Steinberger, for his attention to his duties and kindness to the whole party. Resolved, that we are fully satisfied with our prospects for gold digging, and the general appearance of the country, sufficiently so to make it our permanent home. Resolved, that a copy of the proceedings of this day, together with a small sample of the dust obtained from these diggings, be furnished the *Omaha Times,* and that the editor be requested to publish the same, in order that persons in the states may know how people "out of America," live, move, and have their being.

Resolved, that all papers in Kansas, Nebraska and the several states, who are friendly to the opening of this rich mineral country, be requested to publish the proceedings of this meeting.

Toast and songs being next in order, the following were offered: some of which were received with a hip, hip, hurra, and a tiger.

Regular Toasts

May all the emigration to this El Dorado find as

comfortable quarters as we have in this beautiful camp, and may they succeed in realizing their brightest golden anticipations; hoping that Kansas and Nebraska may not want for good fat beef, no more than we do for fat venison, elk, mountain sheep, antelope, wild turkies, etc. – Gen. Larimer.

The good health and prosperity of Capt. R. A. Spooner – hoping that he may succeed as well in taking back bags of gold dust to Nebraska as he did in conducting our train out. – Hon. William Clancy.

Our cabins and our homes. – Both containing our hearts dearest treasures; the former peopled with pictures of our fondest imaginations; the latter by the reality from which we are separated. – James Kimes.

SONG – The Star Spangled Banner.

The day we celebrate. – May we never be less able to celebrate it than at present, and may we enjoy many a happy return of it. – Volunteer.

Women and wine. – May they both attain that which ruins the one and improves the other; viz: old age. – B. Franklin.

Our homes and those we left behind. – May our toil and industry repay us for the parting and separation. – B. Franklin.

SONG – The Girl I Left Behind Me.

Doctors and ducks. – May the quack of the former be as harmless and of as little use as that of the latter. – Volunteer.

The miners and the mines. – May the latter be as prolific of treasures as the former are pregnant with high hopes. – Volunteer.

The carpenters. – Although not blessed with the presence of women, may they, in the spring find a lucrative employment in the manufacture of cradles. – B. Franklin.

SONG. – Rosalie, the Prairie Flower.

Taos. (pro. Touse) and its productions. – Although there are good things come up from Old Taos,

Its whiskey ain't worth three skips of a louse. – Volunteer.

The past, the present, and the future. – Let the first ever be in remembrance; a bright look for the second; and hope be our guiding star for the third. – B. Franklin.

The new territory of Colona. – May she soon realize her brightest anticipations, and take her place in the galaxy of stars, as a state. – Mr. Blake.

SONG. – The Home of My Boyhood.

Our destiny. – Westward the star of empire takes her way, and she has now lodged in the Rocky mountains, where the original clans see (Clancy) the clouds lowring (Lowry) o'er the peaks of Laramie (Gen. Larimer). – S. S. Curtis.

Direct communication. – May the opening of spring, give us direct communication with our friends in "America," by a regular mail; and although we all acknowledge the benefits derived from our present FEE MAIL, (50 cents per letter,) may we be blessed with an abundance of the genuine article of genus female during the coming summer. – A. O. McGrew.

The press. – That mighty engine which controlls powers and principalities, converts the howling wilderness into smiling fields and busy marts of commerce, sheds its blessings alike upon the rich and poor, the great and the small, the lowly and the exalted; the lever which moves the world. May its influence never be perverted to serve base purposes; may our CASE not be to COPY after others, but may we make it RULE to STICK to our SHEETS (when we get one) as long as there are quoins (coins) in the bank, after which we will DOWN

WITH THE DUST, even though IMPOSING STONES rear themselves before our FORMS in our arduous CHASE after the precious METAL. – A. O. McGrew.

That last toast was received with three times three, and a tiger, after which was sung the beautiful song, entitled, "The mountain boy's call." Gen. Larimer being called upon for a speech, then rose, and made some very appropriate remarks, . . .

A Hit at the Times.

Way out upon the Platte, near Pike's Peak we were told,
There by a little digging, we could get a pile of gold.
So we bundled up our duds, resolved at least to try
And tempt old Madam Fortune, root hog, or die. – Chorus.

So we traveled across the country, and we got upon the ground,
But cold weather was ahead, the first thing we found.
We built our shanties on the ground, resolved in spring to try.
To gather up the dust and slugs, root hog, or die. . .

[Twelve more stanzas]

Adjourning to Auraria, we found the town alive with an influx of miners, some of whom were dressed in the most fantastic and grotesque manner that an active imagination, and the application of the skins of wild beasts could possibly devise. In a short time an immense fire was blazing in the public square, and Terpsichore answered to the voice of Orpheus. Light hearts, merry countenances, and active feet were soon in motion, and the dance continued until midnight. Beneath many a rough exterior, were hearts that throbbed with pleasant thoughts of home. Groups of Indians, with their squaws and papooses, filled-up the background. It was a picture that Rembrandt would have contemplated with delight. On Monday the Masons held their annual celebration in honor of St. John's day, and in the evening had a fine supper; as

I am not so fortunate as to belong to the fraternity, I was not present. . .

You may doubtless hear many reports in the states; some of which are doubtless true, while others are false. Many come here who, because they cannot make independent fortunes in a day, leave, and curse the country. Let them go; we have no use for them here; they are much better at home, where they can have some one to wash their faces every morning, and see that they do not stray too far from home. We are satisfied with our prospects here, and intend to stay here until the whole country is explored. . . A. O. McGrew.

[P. T. Bassett] [247]

Denver City, January 1st, 1859.

Friend Wilcox: – I wish you a Happy New Year! As the express leaves here the first of every month for Fort Laramie, I will write you, to let you know how things are moving.

First – The weather is fine as can be found; I have not had a coat on but a few days since I have been here. We have had about two inches of snow here, and in all, it is the finest climate, so far, that I ever saw for winter; but we can stand and look upon snow four feet deep, fifteen miles from where I write, on the mountains.

I can assure you that what you read is no exaggeration, for it is what I see myself. But what I have done I will tell you. I have some days taken out myself from $5 to $10 with a rocker; but it is hard work, for there is no water to work with. . . The town of Denver City is at the mouth of Cherry creek. Provisions are very high and scarce. I suppose you want to know what the

---

[247] *Missouri Democrat*, February 16, 1859. Bassett was from St. Louis. He was shot and killed by John Scudder on April 16, 1859. See previous footnote on Scudder.

prospects are for trading. I can not advise you, for I think that everybody has written home to their friends to come and bring something to sell. You can judge for yourself how much and many goods will come from the states, and then add to this amount one-third from New Mexico.

I live in a log cabin eighteen feet square, which I built myself, and am my own cook. . . Truly your friend, P. T. BASSETT.

[WILLIAM LARIMER] [248]
Denver City, Jan. 2, 1856 [1859].

DEAR TIMES: – The express started yesterday, but I take the opportunity of a private conveyance, to send you a short letter concerning matters here. . .

I send you some samples of gold dust, and a map of Denver City. We are going rapidly ahead. There is plenty of cottonwood at hand, and pine within a distance of ten or eleven miles. We all build with cottonwood logs, covering the structure with poles, and afterwards with dirt. Larimer street will be the first settled – McGee [McGaa] and Lawrence next; all are "right side up with care." Lawrence, Dorset, William [H.H.] Larimer and myself, all live together at present.

A young man by the name of John Boucher, from Oskaloosa, Kansas, came out in our party. He was taken sick on the way, and lingered till yesterday morning, when he died. Every attention was given him upon the road, and while here, but without avail.

We have good reports every day from the mines, about eight miles from here, on Vaskay's [Vasquez] or Long's creek [present Clear creek]. The miners are taking out from twenty-five to fifty cents per pan.

[248] *Lawrence Republican*, February 3, 1859, copied from the *Leavenworth Times*.

The diggings at that point are extensive, and our success as a mining company is ensured.

Flour sells for $20 per sack of 100 pounds, and everything in proportion. I was offered $50 for my gun, for which I paid $20 in Leavenworth. I am not a hunter, and will sell for $60.

I must close abruptly, and leave much unsaid. You will receive other letters from me by express. Yours truly, WM. LARIMER.

[CHARLES DOLLAR] [249]

Auraria, Jan. 4, 1859.

Since Parkinson left for the east, very little has been done in the way of mining, for the weather has been quite cold, although more pleasant than in some of the states. The climate here is a dry cold, and not at all damp as that of Missouri. The ground is frozen two and a half feet deep, preventing successful mining, but there are a few that piddle a little, and make from three to five dollars per day. I saw a man yesterday, who told me that he had found prospects on Ralston creek, which paid him from twenty-five to thirty cents to the pan. If this is true, they are the best mines yet discovered, but some there are, who hope for still better. Our party has not prospected any yet, but have been busy making town sites and building houses. All in fine spirits and health. I try my luck in the mines in early spring. I tell you we have had to do some hard work here. Harris and I have been in the timber cutting house-logs for some weeks. Money is scarce in Platte valley at present. I have had only one dollar since my advent in this beautiful spot.

There are two stores and three Indian trading posts

---

[249] *Missouri Democrat*, February 25, 1859.

in Auraria; a principal part of the merchandize being whiskey, flour and bacon. They have also a good supply of alcohol, from which is manufactured various kinds of liquor.

I understand from some mountain men that the Indians say they intend to make the whites use all the logs they have cut, in building forts to protect themselves, before spring; but I apprehend no danger. The following tribes vow hostility: Cheyennes, Kiowas and Arrapahoes. I remain yours, CHARLES DOLLAR.

[W. W. SPALDING] [250]

Pike's Peak City, Jan. 4, 1859.

Editor Journal of Commerce: . . . When I say the miners are doing well, I mean that they are making about $5.00 per day; for if a miner starts out prospecting, and cannot average $5.00 per day, he says he has done nothing. But the largest strike that has been made by one man in this part of the mines was $75 in one day, with pick and pan. The lucky man was J. W. Stanley, from Mineral Point, Wisconsin. The largest nugget that I have seen, weighed just $11.10, but the gold in this part of the country is very fine, scale gold.

The Indians are very friendly, so far. Provisions are getting rather scarce, and command a good price. Flour 30 cents per pound, bacon 25, coffee 30, sugar 30, beans 20, rice 30, dried apples 50, tobacco $1.25, whiskey $18 per gallon, with $2.50 worth of tobacco and strychnine in each gallon.

This place sent a man the 5th of December to Laramie, for our mail matter. He brought eight of the *Daily Journals of Commerce,* and 6 *Missouri Republicans,* which were readily sold for $1.25 each number. The carrier could have sold 150 at the same price. If the people in the states took such an interest in newspapers as they do here, there would be no delinquent subscribers.

There was quite an exciting time in our little city, on December 25th. The Indians found out that the whites were going to have holidays, and they came in to join in their favorite holiday's occupation, horse racing. They drove in one hundred ponies to bet on racing. They have a pony that they offer to bet one hundred and fifty ponies, that they can beat any horse one-fourth of a mile. But they got no race. They have a Mexican mule, that they challenge any mule trotting any distance, for one hundred and fifty ponies. But as we had no mules that were trotters, they proposed to trot against our American horses. A Mr. Devine, of our town had a horse that was a good trotter, and with great caution, made a race for eight of the ponies, and won them very easy. Those Indians can be beat, but cannot be backed out on a race. There are two gentlemen of this place going to start for Kansas City or Westport, where they know of a trotting mule, which they say can beat the Indians quite easily. If they can get it, and they can bet the Indians' mule, they can win from two to three hundred ponies.

The health of the country is good. The miners are making great preparations for making their piles in the spring, before the emigration gets here. The all-prevailing law in this country, is the old California miners' law, which every criminal yields to with "high" submission.

I have heard of no disappointed gold seekers in this part of the country. All think their greatest expectations will be realized in the mines. . . Respectfully, W. W. SPALDING.

[JOHN I. PRICE] [251]

El Paso, K.T., Jan. 5, 1859.

Mr. M. P. McCarty – Dear Sir – I write you, supposing that you would like to hear the news from the gold mines, although I have nothing new to say but what you have already heard. There has been no new discoveries made of any importance except some new diggings on Long's creek, but it has been so cold that we have not been able to see how much they would pay, nor am I or any one else capable of giving you a fair representation of the gold prospect here until next spring and summer. There is undoubtedly plenty of gold here, for you can find it nearly everywhere you will dig, but in such small particles that it will only pay where there is water sufficient to run a sluice box. If you come out next spring bring some quicksilver, and a set of carpenter's tools if you can get them cheap. Yours, respectfully, JOHN I. PRICE.

[THOMAS L. GOLDEN] [252]

Arapahoe City, K.T., January 6, 1859.

I am in the gold country near Cherry creek, and I assure you there is gold here, and plenty of it. I have discovered mines that will pay $20 per day; I have not mined much yet, on account of the weather being too cold to work, but am preparing sluice lumber to commence operations early in the spring.

S. S. C. [Samuel S. Curtis], formerly of your city, is here. We have a claim in partnership, one of the

[251] *Ibid.,* March 4, 1859. El Paso was at or near the site of present Colorado Springs.

[252] Mr. Golden was one of the founders of Arapahoe City, about two miles east of present Golden. The *Rocky Mountain News* of October 3, 1860 tells of the marriage of Thomas L. Golden to a Miss Fletcher of Nevada City on September 24, 1860. It says "he is an 'extra clever' fellow." The letter reproduced here is from the *Missouri Republican,* February 8, 1859.

best that has been discovered. Game is in great abundance. I have killed fifty-four deer, six mountain sheep, and a large panther.

The Indians are thick here. We apprehend danger from them. They have sent us word by some of their chiefs, to quit their country, but we think we can stand them a rub, as we have 700 white men here. We have laid out a town by the name of Arapahoe City, after the aborigines. We have also laid off a mining precinct, ten miles square, of which I have been elected recorder. . . Yours, etc., THOS. S. [L.] GOLDEN.

[CHARLES ———] [253]

Auroria City, January 6th, 1859.

A Merry Christmas and Happy New Year to you, friend William. This is all I can pen or send you, at the present stage of the game. Far away from city or home, in the extreme corner of Kansas territory, only fifteen miles from the Rocky mountains, where I now sit in a rough log cabin, over head and around me covered with mud, the floor is made of the same material which is so common in the Missouri river. One side of our house is ornamented by a double set of bunks in which is placed a double armful of grass, and on that is placed a buffalo robe, on which we resign ourselves to the arms of Morpheus. From eight P.M. to daylight the walls of our cabin are decorated with the paraphernalia of an experienced miner. One side of our cabin is covered with fresh venison, for on this article we principally subsist. – Harris started on a hunt Tuesday in order to get fresh supplies of the above article. When we want venison in this country we make up a party, and start for the mountains. Here

[253] *Jefferson Inquirer*, February 26, 1859.

we find a species of black tail deer, mountain sheep and various other animals.

Now for a description of our city, it is situated on the South Platte, at the mouth of Cherry creek, about seventy miles from Pike's Peak. It has about one hundred and thirty houses, and an average of four men to the cabin. The female part is composed of four white ladies, two of whom are married, and the other two single, there are also squaws here, who are the wives of some mountain men, who have located here. When I started for this place, I expected to find but little, or no whiskey, and but few people, but I was much astonished when I found that there were about one thousand persons in the valley, and about ten barrels of whiskey. I am proud to say that Harris and Scudder have thrown off on that game and turned to be sober men. But James B. Reid still takes a nip when he feels an inclination.

Friend Williams. Now for the outlines for my future prospects. I have gone into city property quite largely, I own shares in four town sights. Some of which are, I think, sure to make something for me. I also have an interest in three water claims. These I am told by old and experienced miners, are a fortune, that is, if we go ahead and improve them, which I think we shall. Reid & Co., for such we style ourselves, bought another share in this place; with it is connected the ferry privilege across the South Platte. This, I think is, or will be a grand speculation, that is, if this ever makes a country which I think it is bound to do. Then we are all right. We have as yet done no mining, the weather is too inclement to do anything in the way of digging or washing, the only which has been done was done with a rocker or pan.

There are some men working who are making from five to six dollars per day – a man may go and take a pan of dirt from any place and he will get the culler, that is from three to twenty specks. There are rumors afloat that some places have been found where the dirt pays from thirty to thirty-five cents to the pan. I am of opinion that the best diggings will be found in the mountains. This gold is bound to have a head, and as soon as spring opens our party intend to go and prospect around these mountains a little. We all expect a large emigration here in the spring, all the people here now seem to be in fine spirits, and think they can make a nice thing of it before they go back. Well, I must close, all the boys send their regards. Give my respects to all. Yours, CHARLES ————.

[JOHN W. JONES] [254]

Denver City, mouth of Cherry creek, Jan. 7, 1859

As Mr. Charles Lawrence, one of the commissioners of Arapahoe county, leaves here on to-morrow morning for St. Louis, I embrace the opportunity of writing to you, giving you full particulars of the gold mines and vicinity. The weather here has been very fine, and is now as pleasant and warm as when you left us. The river has not frozen over yet; neither has the soil been froze to hinder the miners from operating. The "Rocker [Rooker?] diggings," is still paying from six to ten dollars per day, carrying their pay dirt to the river in barrows and pans, a distance of five hundred yards. The manner of mining here now is just about the same as when you left us, altogether panning. Numbers are preparing sluices, rockers, etc., for next summer, as they are confident that this is a fine mineral country,

254 *Missouri Republican*, February 23, 1859.

and all are certain and sure of reaping rich harvests in this new El Dorado.

This country not only abounds in the precious metal, (gold) recent discoveries have proved that there exists along the base of the mountains, iron and silver ore in abundance. Large veins of different kinds of coal have also been found; two of which I have found myself, not more than twenty miles from this place. Judge Townsley has just arrived from a deer hunt, in the mountains, bringing in some very fine specimens of limestone; also some fine hard sandstone, which will prove valuable for building purposes.

Our friend Antoine Janiss and his brother Nicholas are camped on Thompson's, along side of a large quarry of plaster of paris, very rich indeed. Also a most beautiful white marble quarry has recently been discovered, specimens of which I have in my possession. There is also mineral salt or alum in great abundance on Vasques fork, ten miles from this place; the miners here use it in making bread, and prefer it all the time to the saleratus, yeast powders, and a host of other poisonous ashes that find their way out to this section of the country. Let the alkali manufacturers keep their trash at home, we want none of it out here, for the simple reason we have found the giniwine harticle here in "de mountain dat beats all competition."

Our agricultural resources I need not dwell upon, suffice to say that not a bottom on the Platte valley from Cache-a-la-poudre to the canon of the Platte river but what is taken and occupied for the purpose of farming next season. . .

In the last two months this has grown to be an almighty fast country, owing to the number of fast men that have emigrated here. We have preaching here every Sunday morning by an "old gentleman miner," of the Methodist race, horse racing after preaching, gambling after speeches, dancing after gambling, and to wind up the grand performances of the Sabbath, after the dance our beloved and most respected friend, the original John S. Smith breaks his squaw's back with a creepy, or a three legged stool, for daring without his permission to trip the light fantastic toe. The original John S. Smith has just been served notice to leave this town in four days; his conduct has been such of late that the mining community will stand it no longer. His squaw is suffering severely from the effects of the said creepy. The original John S. goes into exile alone, among the Arapahoes, there to recruit his almost shattered fortunes. Two trains have just arrived from New Mexico with dry goods, provisions, etc. Flour is selling here at ten and twelve dollars per cwt.; bacon at 25 cents, and everything in proportion, very reasonable. Onions, potatoes, cabbages, etc. are pouring in here from Mexico, so instead of passing a severe winter here, as we expected when we left Laramie, it is quite the reverse, for we have very comfortable houses, and living on all the luxuries that New Mexico affords. By the by, our Christmas and New Years passed off very pleasant, and I can safely say that I have not had a more real happy social time in fifteen years than I passed here on Christmas and New Years with my friends.

I send you in this letter some specimens of gold dust that I panned out of the bank of the Platte river on yesterday; the amount is one dollar and a half ($1.50), just the work of twenty-three minutes to dig, pan out, and clean, ready to send to a friend a specimen of Platte valley gold. Whenever I write to a friend and wish to send him a specimen, my little squaw [255] and myself take a pick, spade and pan, and down we go to the bank of the river, not more than five hundred yards from my house, we pitch right into any place along the bank, returning in half an hour, at farthest, with specimens of one, two, and sometimes, three dollars, (for the truth of this I stake my honor, in case you want to publish it.) Further, an Arapahoe Indian named *Cut Nose* arrived here two days ago, having with him some shot gold, to the amount of fifty dollars. It is a pretty fair specimen, from the size of a pea to a large buck shot. He says that he found it recently in a crevasse in the mountain, while hunting elk and mountain sheep. . .

Since I have written the above, Judge H. P. A. Smith has called in, by way a most peculiar friend of mine, and wishes me to tell you, the people, and government, that unless they make a treaty with the Indians here early in the spring, we will unquestionably have trouble. This is true, and in fact our chief reliance is upon the government to aid us out of these troubles and difficulties that we may expect to have.

Dr. Phip, I am putting things right through here; for instance, two ferry boats under way, and five good houses up by the time you get here, all finished and in good order. I am putting you up a store 25 by 40 feet. You will have no trouble but unload your wagons and pile up on your shelves. The ice house is under way, and will soon be finished, and if ever this tedious river

255 This mention of his squaw leads me to surmise that the writer of this letter is William McGaa, alias Jack Jones, prominent Indian trader at pioneer Denver. He was one of the original directors of the Denver Town Company, 1858-1859. The original record book (in possession of the State Historical Society of Colorado) gives his name both as Jones and McGaa. McGaa claimed to be the son of an English nobleman. He died in the Denver jail in 1868. See J. C. Smiley, *History of Denver*, 221-222.

freezes up, up and in goes 500 tons of ice with a vengeance.

I have a big lay out, the best in this Platte valley, and with your assistance we can make a pile. Enclosed I send your certificate of stock in Denver City. You may need it. It is legal, and signed by the vice president, which is myself,[256] the president being absent; our constitution provides that the vice shall act in his place. It is also signed by Mr. Charles Lawrence, who is our secretary. It is the best property in this country, and lots are now selling at $25 and $30. Mr. Lawrence will call and see you; he will give you all particulars. I rather think you will find a ready sale for some of it; it will be worth $100 a lot next summer. JOHN W. JONES.

[J. H. MING] [257]

In camp, foot of Pike's Peak, January 7th, 1859.

Jas. M. Ming – Dear Brother: A gentleman has just come into camp, who is on his way to Fort Laramie, and I embrace the opportunity to write you a line by the light of a pine knot. I am now encamped on the divide, between the Platte and Arkansas rivers, and on my way to a little town on the Platte, where I expect to open my stock of goods. I have three wagons loaded, having left the other three at Fountain City [present Pueblo], where we arrived some three weeks ago, built a house fifty feet long and eighteen feet wide, opened some goods and commenced the trade. There being no other store in town we had a big run. Did pretty well for awhile but the money soon gave out

---

256 The original record book of the Denver Town Company is incomplete. It does not name the vice president of the company. But it is entirely likely that Jones (McGaa) held this office in January 1859.

257 *Hannibal Messenger*, February 18, 1859, copied from the *Washington* (Missouri) *Advertiser*. The Ming and Cooper store was one of the first in Denver.

and we had to trade in stock – horses, mules and cattle. Cooper remains there, while I go to the Platte, hoping to sell out by spring. We can get but little money. Could sell any quantity of goods on credit, but it won't do. We expect to trade almost altogether in stock. We have also, considerable Indian trade.

The weather has generally been very good, and our stock is getting fat. Last night was very cold. The snow is six inches deep, and the thermometer stood at fifteen degrees below zero at daybreak, but it is more pleasant now. We are now encamped at the place where men and animals froze to death last May.[258] The grave of one is near. It is considered a very hazardous undertaking to cross the divide at this time of year, notwithstanding I am trying it.

Some of the boys are with me, and some are up in the mountains prospecting. All are well and fat. I know nothing of the mines yet. I have not much faith in them, that is, as far as they are known. I think nothing is known yet to induce persons to come here to mine, but it is my opinion, as well as many others, that good diggings will sooner or later be found here. There is very little doing over in the mines, but if I am lucky I will be there in two or three days, when I will tell you all about them.

I am undetermined as to what I shall do about coming home in the spring or fall – all depends on circumstances. I want to make some money first, and I am trying hard. J. H. MING.

[A. O. McGREW] [259]

Auraria, Jan. 10, 1859.

Presuming that your readers would like to hear from

---

258 Marcy's expedition. See the preceding volume in this series, page 105.
259 *Omaha Times,* February 17, 1859.

this country, its prospects, etc., I will dot you a few items, giving you the truth about gold as far as come under my observation. Men who are mining at the Spanish diggings, four miles above here, on the South Platte, average from three to five dollars a day with rockers, carrying the water from the Platte, some fifty rods. The better way for your readers to judge of the yield of the mines, is to know the "prospect of the pan." I have seen 50 cents in a single pan, a few times – ten to fifteen cents frequently, and hardly ever less than five to eight cents. Californians say, that when they get sluices arranged, they can make five to fifteen dollars every day they work. Companies are in the mountains above and below here, prospecting and some of their men have been here, who report that they have discovered coarse gold.

Capt. [Marshall] Cook, with a party of some 30 or 40 men, are camped on Long's creek [Clear creek], some 12 miles from here, and have prospected enough to satisfy themselves that they can take out from ten to twenty dollars per day. They are erecting two saw mills on the above-named creek, and preparing sloughs and sluices to commence operation early in the spring. Capt. Aiken,[260] and a large party of men are on Bascon's creek [Boulder], some 40 miles below here, in the mountains, and have discovered shot and quartz gold. Capt. Aiken is an old Californian, and an energetic man, and he is convinced that the mines in this country will pay better than the best mines in California. Capt. [S.S.] Curtis and others from Council Bluffs, are out on Long's creek – have laid off a town [Arapahoe City], and are making preparations to mine as soon as the weather will permit. Gold has been dis-

260 Captain Thomas Aikens was one of the founders of Boulder, Colorado.

covered on Cashlapood [Cache la Poudre] creek, which is 60 miles north of this place. . .

[EDWARD COOK] [261]

Thompson's fork, Jan. 10, 1859.

I should have written to you before, but our company has been prospecting the country for the last five weeks. We have prospected nearly one hundred miles along the mountains, and have found fine gold on nearly every stream. . .

Cherry creek is forty miles from here. At its mouth is a new town called Auraria, containing fifty houses, finished, and forty or fifty in process of erection.

Most of the citizens are intelligent and enterprising young men, of whom there are about four hundred — and three women. The number of men in the country is estimated at twelve hundred. Have had but little snow yet; our cattle are doing well.

I do not advise any one to come here; I tell them what I know and they must judge for themselves. One thing I will say, I am not sorry I came, and expect to make money as soon as spring opens.

As soon as the weather will permit people to go into the mountains, I expect some big strikes will be made. Every one seems to think he has a good thing and is determined to keep it. If you think of coming let me know as soon as possible. Yours, etc. ED COOK.

[SAMUEL R. STUMBO] [262]

Rocky mountains, Boulder City, Jan. 11, 1859

. . . We arrived within 15 miles of Cherry creek

----

261 *Kansas Press,* February 19, 1859.

262 From the *Falls City* (Nebraska) *Broad Axe,* and reprinted in the White Cloud *Kansas Chief* of March 24, 1859. At the close of the letter the editor says that Mr. Stumbo and a Mr. Yount have just returned home.

on the last day of October, and found it all humbug as to the plentitude of gold. That there is gold all over this country, there is no doubt, but there has nothing yet been discovered that will make a man stay here.

Miners are making from 10 cents to $1 per day. That is what our men are making, and it is said that we have the richest diggings that have yet been discovered. Some men will say that they are making from $5 to $20 per day, but they never can show it. The largest specimen I have got yet is worth about 20 cents, and I believe it is the largest that has been found. We are in hopes that something better will be found in the mountains, and have sent five men out to prospect. If they do not find a better prospect than we have now, Mr. Yount will go home, and I think Lewis Davenport and myself will go with him, for a man can't make his board here at mining.

The above is about all the news in relation to mining that is worth giving. The reports you see in the papers, I warn you not to believe, for they are put in circulation by town builders for speculative purposes, and to them may all the excitement be traced. There is nothing here to pay for coming.

I have got just about gold enough for a ring, which, if I do not come home myself, I will send you by Mr. Yount. Your son, SAMUEL R. STUMBO.

[WILLIAM LARIMER] [263]

Denver City, Arapahoe co., K.T., Jan. 15, 1859

[A long letter discussing routes to the mines, a railroad through South pass, the good prospects in the mines, food supplies, etc. The letter ends thus:] Claims are being taken here all over the country, more particularly in the choice spots on Cherry creek, and up

[263] *Leavenworth Times,* February 12, 1859.

and down the South Platte. We have regular claim clubs, and miners clubs, with Methodist preaching every two weeks by Rev. G. W. Fisher, of Oskaloosa, Kansas. We have also Saunders & Co.'s express, once a month to Ft. Laramie. Cost of letters, 25 cts, single letters, 50 cts, and 15 cts, for newspapers. Smith's express for Leavenworth, left here on the 20th of November and he charged one dollar per letter. We expect his express back about the 20th of this month. The people here have all supplies of provisions to last until the first or middle of May – perhaps longer. WM. LARIMER, JR.

[JOHN SCUDDER] [264]

Auraria, mouth of Cherry creek, Jan. 16, 1859

. . . The population of the valley is said to number two thousand souls, but I think the number over-estimated. There are about thirty St. Louis men in this place. Our camp consists of A. G. Baber, a son of the former auditor of the state; O. B. Totten, John Harris, Charles Dahler, Charles Noble, James B. Reid of Glasgow, and myself. We think we have got a good thing, and expect to get rich out of it. We have linked our fortunes with the Georgia company, who first opened the mines in this country in the spring of 1858, and have laid out a city named as above – *Auraria*. It is beautifully situated at the mouth of Cherry creek, on the south bank of the South Platte. The mountains are about twelve miles from us, but to the eye it looks to be only about two or three miles. We have over two hundred good log cabins built, and a population of about five hundred men and three ladies. I think we will be able to furnish accommodations for five or six thousand people in the spring. Every man who can

---

[264] *Missouri Republican*, March 5, 1859.

get a team and axes is hard at work building cabins. The lots are large – 66 by 132 – and the place already looks like a city. We have a claim club, signed by half the men in the valley, whose business it is to see that all town sites and farming and timber claims are recorded. So far we have done well, and hope the county officers that will be elected in March will take up the business as we leave it. If so, there will be no trouble, if not, the bloody scenes and riots of eastern Kansas will again be enacted, and many good men will fall in the defence of their actual rights.[265]

One third of the population is composed of broken down land speculators from Iowa, Nebraska and Kansas, and they are ready for anything that will make trouble.

Gov. Denver sent out last fall a body of men with commissions sufficient to organize and carry on the business of the country. They have not all done their duty, and the people are not satisfied with their conduct. The people had a meeting in the last of December, and passed resolutions expressing their dissatisfaction.

An election will be called for the 4th of March, and already the ball has been rolling to bring in men from Arkansas and Nebraska, making this county over two hundred and fifty miles long, and heaven knows how wide, for the territory west of the mountains extends into the eastern line of California. . .

We look for a large emigration in the spring. . .

We all send love and respects to our friends; tell all who come to the country from St. Louis, to look for the St. Louis ranch, which may be found on Ferry street, about three squares from the river in front of which will be found a wagon-box on end, which reads

---

[265] This probably refers to the friction between the promoters of Auraria and of Denver. The animosity led to the killing of Bassett (an officer of the Denver Town Company) by Scudder, as related previously.

thus: Office of John Scudder, recorder of claims, land
agent, houses for sale and to rent, pine lumber for sale.
Walk in – office hours when the door is open.

Henry Brittingham, a nephew of Mr. Crow, of the
firm of Crow, McCreery & Co., died Wednesday,
January 12th. He was buried by strangers on the bank
of the South Platte. . .

Yesterday a man who was caught stealing from a
merchant, was tried by seven men, and found guilty.
He received thirty lashes on the bare back, and was
ordered to leave the district to-day, never to return
again. JOHN SCUDDER

[JOHN L. HIFFNER] [266]

Auraria City, K.T., Jan. 17th, 1859.

Mr. Wm. E. Collins: – Sir: You are aware that I
left for Arizona. I found when I got to Santa Fe, that
I could not get any further with so small a party; we
sold out at said place, and myself and three others
started for Arizona, and oh, what a time we had. When
about four hundred miles from Santa Fe, we were at-
tacked by about ninety Indians – they took our wagon
and stock after about one hour's fight, and followed
us for 240 miles, wounding one man slightly, and another fatally. You may think strange that I am here,
but the above statement will satisfy you that I could
go no further, so I turned my course for this place. I
landed here the last day of December, 1858. When I
got here I found about 600 men, mostly engaged in
building winter quarters. I have not prospected any
yet, though I have seen a great many specimens. The
winter is too cold to work in the mines, though some
few are at it, making from $3 to $12 per day. Nearly

---

[266] Kansas City *Journal of Commerce*, February 23, 1859; reprinted from
the *Liberty Tribune*.

everyone has a claim picked out to work as soon as
spring opens; and they say they can make from $8 to
$50 per day. I will have an interest in a mining com-
pany as soon as they can be worked; at present I am
engaged in building. I have an interest in the aforesaid
town. There is some 139 houses at this time, and from
3 to 4 going up every week. This town is at the mouth
of Cherry creek. I saw a man yesterday wash three
buckets of dirt and get eight dollars and ten cents in
gold. JNO. L. HIFFNER.

[H. L. BOLTON] [267]

Auraria, Jan. 19, 1859.

All the accounts of gold findings of an extravagant
character are the fabrications of speculators. I wish
to put you and others on their guard against these
stories; especially Gen. Larimer's account. I will venture to say that he does not know anything about the
matter. I have not found a good prospect yet; I am
on the ground. I venture the prediction that few persons will make fortunes hunting gold in this country.
But as "seeing is believing" let all who wish to have
a sight of the "elephant" come on. I am beginning to
get a view of him.

There are more than 200 cabins built here and 200
more are to be erected before the last of March. A
good hotel will be ready for "boarders" by the end of
May. It is to be two stories high; 75 ft. in width and
120 ft. long. Speculators are already busily engaged

[267] *Chicago Press and Tribune,* February 21, 1859, reprinting from the
St. Louis Democrat of February 17. The unfavorable report of Bolton is
followed by letters from John Kearns, John Bruce, and J. G. W. Coonce
which are more favorable.

Charles F. Connelly wrote a long letter from Cherry creek on January 22,
1859 (published in the *Council Bluffs Bugle,* March 16, 1859). It contains no
new information of importance.

in laying off cities around the diggings and they are the fellows who are sending to the states such glowing accounts of gold discoveries. . .

The principal amusement here during the winter has been card playing, telling yarns and drinking most execrable whisky. The latter is worth $10 per gal.; in St. Louis it would cost 20c.

I must not omit to tell you that I have not seen a white woman since I left the states. H. L. BOLTON.

[RUFUS C. CABLE] [268]

South Platte, January 28, 1859.

Dear Brother: . . . There have been no mines discovered yet which are worth working. Any place here one can get small quantities, but will not pay more than $2 or $3 per day, and the ground is frozen so that no person tries to make anything. When spring opens I think we will find good mines. . .

Yesterday I went to Auraria, to witness a feast given by the whites to the Arapahoe Indians. There were assembled some 500 warriors, and 300 or 400 women and children. Two oxen were roasted, and piles of dried apples heaped up on blankets, coffee, bread, etc. I have never seen men eat till now. I have heard that one man could eat an antelope at one meal, and I verily believe one Indian, called, "Heap of Whips," could eat a whole ox. It is worth one year's travel to witness such a scene.

When I come from the Cassa la Poudre, which will be in ten or twelve days, I will write you.

I have sixteen lots in El Paso, at the base of Pike's Peak; and if I do not find any gold, I shall go there and build houses. . .

---

[268] Kansas City *Journal of Commerce,* April 6, 1859; reprinted from the *Western Argus.*

Some of the boys have frozen their feet and have all their toes cut off. But, so far, I am safe, and enjoy better health than ever I did in my life, and am not at all disheartened, for I think I will make money here by next fall, if I live.

We are expecting 500 Mexicans here in two weeks, and am cultivating my Spanish. Yours, ever, RUFUS E. CABLE.

[A. A. BROOKFIELD] [269]

St. Vrain's creek, Rocky mts., January 26, 1859

Friend Norton: – Yours of December 16th was received this evening, and in conformity with your request, I immediately sit down to reply. And firstly, if you have any confidence in judgement, and act on my advice, you will immediately abandon the idea of coming out here, without some other discoveries are made than have been thus far. My impression of the mines is that they are a d—d humbug. I wish that I could write otherwise, and particularly on my own account, as I should be very glad to see you and Mrs. N. here, and it would be such a good opportunity for my wife to come out. I have already written her that if I stayed here I should send for her.

Where we are wintering is about twenty miles north of the mouth of Cherry creek, and in a canon at the foot of the Rocky mountains, or Black hills. We have found here the best quality of gold that has been discovered, and cannot make one dollar per day. Billy Moore, Tom, and myself are working together; we have built a good dam across the creek, and have our long tom set (the only long tom that is set and worked

---

269 Kansas City *Journal of Commerce,* March 16, 1859, and reprinted in the *Herald of Freedom,* April 9, 1859. The latter paper gives the date of the letter as January 26. Brookfield was ex-mayor of Nebraska City. He was one of the founders of Boulder, Colorado.

in the country), it works beautifully, so the old Californians say. There are six old miners in our settlement. I don't think that there is another company that came here last fall who have done more work or more extensive prospecting than we. We have prospected for some twenty miles south of the mouth of Cherry creek to the neighborhood of Long's Peak, north. Besides, we have prospected as far into the mountains as we could get on account of snow.

The only hope that I have now, is that when spring comes, we may find something to pay us in the mountains. Should we be so fortunate, I will immediately write you.

I suppose you have, ere this, learned that Mart has gone back, also E. Muir, Stuffe, Faith and Dr. Mathews. This I send by Capt. Young of Archer, N.T. Jas. Aiken, Sam'l Aiken (Rockport company), L. Davenport and S. Stamben, of St. Stephens, N.T., leave for home in a few days; and were it not for the dread of cold weather and the hope that we may discover something in the spring, I have no doubt but that nearly the whole camp would leave for home.

We were quite surprised a few days since when we read the glowing account in the Missouri river papers, of what the miners are doing out here. I pronounce them a pack of lies, written and reported back by a set of petty one horse town speculators, and are calculated to ruin many a poor devil besides your humble servant.

That there may be gold discoveries to pay, I will not deny – hope it will. I am here, and may as well stay until satisfied. Should anything turn up, I will inform you, and shall depend on my wife coming with yours, otherwise you may expect to see me on the Big Muddy by the 14th of July, certain.

Accept my best wishes to yourself, and believe me,
Yours truly, BROOKFIELD.

[JOHN POISAL] [270]
Auraria City, Jan. 30, 1859.

There are at this time about one thousand men in
the mines. Some are at work, and others doing nothing.
Those who work make from five to six dollars per day,
working o[nly ha]lf the day. It is the general opinion
among our citizens when the spring opens very rich
gold mines will be discovered. The gold already found
is of good quality, and everybody seems well satisfied
with the prospect.

I want you to come out early in the spring – a fortune
is in store for us. There is gold here in considerable
quantity, and no mistake. Bring picks and shovels, groceries and provisions, for they are in demand, and sell
at high figures.

[M. S. WILDER] [271]
Pike's Peak, Feb. 1st, 1859

My Dear Brother: According to agreement when
I left home I send you the following news:

I have been mining for the last four weeks, and my
average is from two to five dollars per day. It is hard
work, but pays well, if one will only keep at it. My
living is very good for a new place like this – plenty

----

[270] Extract of a letter published in the *Kansas Weekly Herald*, March 19,
1859. Poisal was an old Indian trader. The newspaper says he had been in
the country for twenty years. The letter was written to his son-in-law, L. J.
Wilmot. Poisal, who had married an Arapaho, was the father-in-law of
Thomas Fitzpatrick, noted mountain man and Indian agent. After Fitzpatrick's death (1854), his widow married Wilmot. See Hafen and Ghent,
Broken Hand, the Life of Thomas Fitzpatrick.

[271] *Missouri Democrat*, March 15, 1859, quoting from the *Chicago Ledger*
of March 6. An unsigned letter, dated at Auraria, February 1, was published in the *New York Tribune* (weekly), March 12, 1859. It gives a generally favorable account of the country.

of game of all kinds. A portion of the country is good farming land, with timber and water in abundance.

When I left home you thought of coming out in the spring with some goods and liquors. I would advise you to bring out coarse clothing, groceries and provisions, as they will pay best. Liquors of all kinds I would let alone, as there is abundance of them here now, and every wagon that arrives is loaded with them. My opinion is that within three months from now, you can buy liquor as cheap as you can in any of the cities east. . .

I will say one word more, and that is about the weather. It has been a very fine winter so far. I have not worn a coat half the time since I arrived here, which was Dec. 14th, inst. My health has been perfectly good.

I hope you are well at home. Your affectionate brother, M. S. WILDER.

———

[WILLIAM LARIMER] [272]
Denver City., K.T., Feb. 2, 1859.
[He first writes of the Indians encountered in coming up the Arkansas.]
. . . We had a little specimen of Judge Lynch a few days since. A person was charged and found guilty of theft by a regular jury, and he received his thirty lashes – "whipped and cleared." All was done cooly and calmly.

A. W. Bacon, Parker, and three other men, from Pontiac, Michigan, came through by the Smoky Hill, the whole distance. They were delighted with the route. They made their own trail with one wagon; passed through a beautiful route the whole distance, and found no trouble crossing the streams. They left Leav-

[272] *Leavenworth Times,* March 5, 1859.

enworth on the 24th of October, and were caught in your heavy rains. They took their time, often laying up a week at a place. Mr. Bacon speaks highly of that country the whole distance. . .

The country is fast settling up by claims being taken. There is a town laid out every twenty miles clear to Fountain City, a distance of 110 miles. They range from here as follows: Russelville, Point of Rock, and Forest City (this is to be the county seat of Westmoreland county). Pike's Peak is situated in this county, and about twelve miles from Forest City. The next town on the main New Mexico road is Junction City, on the Fountain Quabo, or Boiling Spring. The next is Antabee [Autobee] (named after the old trader.) The next, Fountain City, on the Arkansas. A court house is under contract in this city (the county seat of Arrapahoe county). A bridge is also under contract, crossing the South Platte river. This bridge will be of great value to our city.

Auraria is immediately opposite us, only separated by Cherry creek. This is quite a town. Auraria is on the west bank of Cherry creek, and Denver on the east. Auraria was commenced first. I kindled the first fire in Denver City, and now we have one hundred and fifty houses built and under way, with at least five good hotels in the number, with buildings of every description. Messrs. Blake and Williams, of Crescent City, Iowa, are building a house 64 by 96 feet. This is intended for a concert room. Our pinery is twenty miles off.

Hundreds are up there getting out boards, shingles and timber. Window glass now sells at 30 cents per pane 8 by 10, and none here; so with every other article for building purposes. Many are putting on their roofs with weight poles for want of nails, and skins for glass.

Everything is wanted here from a needle to an anchor.

Since I commenced writing this letter the mail from Fort Laramie has arrived. It came in yesterday, the 1st. I received a number of your excellent paper dated 25th Dec. (weekly). I have read it clear through, advertisements and all. . . WM. LARIMER.

[C. L. COOPER] [273]

Fountain City, Feb. 3, 1859.

Mr. Editor. – Dear Sir: As I have promised to write to several of my friends and acquaintances, you will very much oblige me by publishing this in your paper.

I have just returned from our trading house at Cherry creek. There is a town there with about two hundred houses, and I think about seventeen hundred persons engaged in building and mining, and all very much pleased with their prospects for a fortune.

There has been a large scope of country prospected, reaching from the mouth of Cherry creek, along the base of the mountains, to the south of Arkansas, a distance of two hundred miles, that raises gold from three to sixteen dollars per man per diem, where the mines are fairly opened and water can be obtained. This gold is fine, but can be gathered by the usual mining apparatus. . . I know that there is gold here, for I get it in exchange for goods, but to what extent I am unable to say.

There is one thing more that I know, and that is, there is the nicest farming valley here that I have yet seen, and capable of producing everything common to this latitude. The climate is beautiful and mild. The ground has not been covered with snow this winter.

---

[273] *Missouri Republican,* April 18, 1859. Ming and Cooper brought trade goods out to the mines. Cooper remained at Fountain City (present Pueblo) while Ming continued to Denver. See Ming's letter of January 7, above.

I am now using hay that I have had cut since Christmas. . .

Fountain City is a considerable town of about fifty houses.

Yours, with respect, C. L. COOPER.

[W. R. REED] [274]
Winter quarters, 2½ miles from Cherry creek,
Feb. 9, 1859

Dear Sir: – . . . The weather is warm. We have had no real cold weather this winter. There is considerable excitement about the gold here. There is some men who work with their long toms and making it pay very well. They can't work only from about 10 to 4 – then the water freezes in the bottom of their toms. . .

The men are preparing to go to work in the spring – they are sawing lumber for their long toms. Saws are very scarce – there is but three whip-saws in the country, and lumber is in great demand.

Mr. Horton, I had the blues for a while, and thought I would go back. I came across one of the Georgia company, and he said that he could at least make $15 per day in the spring, and keep it up all summer. He also said that if I would stay and not make $10 per day, every day I worked, he would pay me a $1000 next fall.

Up in the Spanish diggings there is a Dutchman working with a long tom, and is taking out $10 per day, working only four hours per day, and he says that his claim will turn out from the surface of the ground down ten feet, 15 cts. to the pan, and when the weather will permit him, he can wash out 500 pans per day. . .
Yours truly, W. R. REED

---

[274] *Council Bluffs Bugle,* March 30, 1859.

PACIFIC RAILROAD PROCEEDINGS AT DENVER CITY, K.T.[275]

A large and enthusiastic meeting of the citizens of Arrappahoe county, Kansas territory, was held at the house of Gen. William Larimer, jr., in Denver City, on the 4th of January, 1859, to take into consideration the subject of the Pacific Railroad.

When on motion of Hon. William Clancy, Gen. Larimer was called to the chair and Dr. A. S. Kunkle, A. J. Williams, Marshall Cook, E. P. Stout, James Lowry, Dr. L. J. Russell and W. H. Brannan were appointed vice presidents; Chas. A. Lawrence, secretary, and Samuel S. Curtis, A. Sagendorf, I. T. Davis, T. H. Russell and Richard Whitsell [Whitsett] assistant secretaries.

On motion of D. C. Collian [Collier], the following gentlemen were appointed a committee to draft a preamble and resolutions to be submitted to the meeting: D. C. Collian, Hon. Wm. Clancy, Hon. E. P. Stout, W. M'Gaa, Esq., Capt. P. T. Bassett, Capt. Wm. Parkinson, G. Bushey, Capt. John Scudder, John M. Kirby, C. H. Blake, L. Tierney, Geo. Gildersleeve, Fulsum Forsette, Judson H. Dudley and Capt. W. Spooner. The committee thro' the chairman D. C. Collian submitted the following preamble and resolution, which after being fully discussed by W. H. Braman of Missouri, Hon. Wm. Clancy and E. P. Stout of Nebraska, Dr. L. J. Russell of Georgia, D. C. Collian of Kansas, Capt. Parkinson of St. Louis, Williams and M'Gaa of the mountains, and the president, of Kansas, was unanimously adapted.

WHEREAS: The subject of building a rail road to connect the Atlantic with the . . .

[275] *Nebraska Advertiser*, February 10, 1859.

[W. P. POLLOCK] [276]

Arapahoe City, K.T. Feb. 11th, 1859.

Friend Cundiff: After considerable study, I have once more concluded to write you, concerning the country, but have not been able to learn much of it since my last – the weather having been so bad that we could do nothing in the mines.

A reliable party of some seventeen men, started for the mountains in search of the lumps, and after traveling for sixteen days, in the deep snow, they returned unsuccessful; and some had come to the conclusion that there was no gold in the country – others being satisfied to hold on until the snows got off so they could have a chance to prospect. I was out with the party for six days, and thought it would be more pleasant to stop in the valley, until the weather suited better for the trip. I think there will be heavy gold found as soon as the frost gets out of the ground, and not before. . .

I do not advise those who are in good circumstances at home to come here. Men who are doing well had better stay where they are. . . Respectfully, W. P. POLLOCK.

[CHARLES H. NICKEL] [277]

Denver City, Feb. 12, 1859.

Dear Brother: – Since writing to you last, I have been painfully compelled to change my opinion of the prospects here, as regards gold digging. If this reaches you at the point proposed in my prior letter, heed my words, starvation must stare you in the face, no matter how well you are prepared for the trip. Robbing emi-

[276] *Missouri Statesman,* April 1, 1859; reprinted from the *Holt County News.*

[277] *Atlanta Weekly Intelligencer,* May 11, 1859; reprinted from the *Calhoun Platform.*

grants of their provisions will be the result of those unprepared, and more suffering will be the consequence, than ever has occurred in the history of similar adventures in the United States. That there is some gold here I do not deny, but that it will pay to dig it, by means that has ever been adopted in other mining districts I do assure you and others that it will not. I will return shortly for my old home, to once more enjoy the society of my friends. Yours affectionately, CHARLES H. NICKEL.

[O. P. GOODWIN] [278]
Boulder City, Feb. 12th, 1859.

Gen. Eastin: . . . The Boulder district diggins, as far as prospected are the richest yet discovered. They are situated on the head waters of St. Vrain's fork of Platte, and are given up, by all "old Californians" in the mines, as the richest they have ever worked, paying from 3 to 5 dollars per day to the man. . .

Now a word for the interest of Boulder City, which is situated on the head waters of the St. Vrain's fork of Platte, 30 miles from its mouth, and 3 miles from the foot of the Rocky mountains, in a beautiful valley, with a commanding view of the country. Of course, like new cities, it has its natural advantages, and certainly we have as good a right as anybody to uphold our own interest, and show the fortune-seekers where they can best employ their time when they reach this golden land. One thing certain, it is the head of "wagon navigation" for the mines in the mountains. Here a man can easily pack a mule with an outfit for days, for any portion of the mines he may think proper to visit. It has the best farming land in the country, ad-

[278] *Kansas Weekly Herald,* March 19, 1859. It was copied in the *Hannibal Messenger* of March 22.

joining the best of pine and cottonwood timber. One of our citizens has already a saw and grist mill en route from Nebraska City, and another has gone for goods to the states. The population exceeds 280 male, and but one drawback, there is not a single white woman in the country; but we hope when spring comes, with it will come plenty of help-mates for the isolated miners. Yours respectfully, OLIVER P. GOODWIN.

[D. C. COLLIER] [279]

Denver City, Feb. 12, 1859.

We have been assuming more and more the appearance and habits of civilization. . .

The mania for town building is as great as ever it was in eastern Kansas. The best sites are this place, and Fountain City, at the mouth of Boiling Spring river. Between these points there are three towns – Forrest City, Russellville and Junction City. There is a company formed to run a stage across. There are also Columbus and Eldorada, near the base of Pike's Peak, Arapahoe, at the base of Table mountain, and a host of others which I cannot now name.

Our election for county officers comes off in March. Politics will have no influence in it, but it will probably be a sort of "free pitch in."

Several companies have come through by way of the Smoky Hill. Their representations make it the best route traveled. They report a good supply of wood, water and grass. They found deposits of iron, coal and chalk, all of the best quality and in the greatest abundance. The Indians were extremely friendly, and ever ready to point out the road. . .

Universal good health prevails. I am getting so fleshy that my clothes are quite too small.

------

[279] *Herald of Freedom*, April 9, 1859.

[H. L. BOON] [280]

Denver City, Feb. 12, 1859.

Many reports respecting the South Platte gold mines, have no doubt reached you ere this time; . . . At what is called the Mexican diggings, one mile and a half below the town of Montana, and three miles above the city from where I write, there has been a considerable amount of gold taken out. On Long's creek also rich diggings have been recently discovered, paying the miner at least ten cents to the pan. On Cherry creek, from the crossing, or the foot of the divide to its mouth, fair prospects have been found. On Dry creek, some seven and a half miles above this point, on the Platte, perhaps the richest deposits have been found, but yet the mines have not been successfully worked, owing to the scarcity of water at this season of the year. . .

So far as the country for other purposes is concerned, I dare say there is no better. That the Platte and Arkansas bottoms will yield abundantly to the industrious farmer, there can be no doubt, neither can it be excelled for grazing cattle. . .

The climate is unequalled, we have had a remarkably mild and pleasant winter. . .

There are many old mountaineers and Indian traders here, and it is amusing to them to hear the "young America" talk of real estate, houses, lots, town shares, original interests, etc. Here on this spot, where now you see a flourishing and enterprising town of nearly one thousand inhabitants, with dry goods stores, tin shops, real estate agencies, blacksmith shops, law offices and doctor shops, four months ago was a beautiful valley filled with game of every description. . .

The people here are expecting a heavy emigration next spring, and in anticipation of it, are making all

[280] Kansas City *Journal of Commerce,* March 25, 1859.

the necessary preparations to receive them, in the way of building them houses to live in as well as houses for business. We say, let all come who will. Very truly, HAMP. L. BOON.

N.B. For God's sake send me a copy of your paper, a *Journal of Commerce* is worth $2.50 here. Send by way of Laramie. H.L.B.

———

[GEORGE STEPHEN] [281]

Auraria, South Platte, February 13th, 1859.

Mr. James Somerville: – Having an opportunity, by our newly-appointed postmaster, Mr. Samuel E. [S.] Curtis, returning to the states, I send you this, my second letter from here.

Although we are not without hope of finding gold in the mountains that may pay, yet it is my duty to you and our friends to inform you that the diggings, so far as known, will not pay; they do not yield fifty cents per day to the man, with the best appliances for collecting it in sluices and about ten cents with a cradle. There have been exceptions to this; some small patches payed a few dollars per day; but they were soon worked out.

This gold-field is of vast extent along the foot of these mountains, reaching the breadth of this territory; and beyond, if we can believe reports, the color is found on every water-course, and on the open prairie, in many places; but in general it is so fine and light that it requires much care to collect, and much bulk to the value. A quantity that a company had been working for a week and over, I estimated to be nearly an ounce. It weighed six dollars. Dry it, and cast it

———

[281] *Herald of Freedom,* April 2, 1859; reprinted from the *Johnson County Standard.* In introducing the letter the newspaper says that Stephen has had years of experience in mining, both in Australia and California.

LITTLE DRY CREEK DIGGINGS

on the water, and part of it will float, like chaff, for a time.

When I see a stray newspaper from the states, I find there "plenty of gold found here, and to be found." The writers of those are town lot speculators, and traders, who rely on the emigration alone to favor their schemes. . .

We have had a beautiful winter – clear days and frosty nights. Our cattle are doing well, it being such a winter as we had last, up to this date. James Stocklin and I have but one good yoke of oxen and a wagon, but we keep them. I have been employed most of my time sawing lumber, and digging ditches to sluice with; but as we have more lumber than we want ourselves, and can sell only on credit, we have stopped. I am an expert with that "long saw," and could do well if there were buyers.

Mr. James left us five weeks ago, on a secret expedition, believing some grand discovery had been made, but the result is not known. I have had two letters from him across the "divide" at the base of Pike's Peak. They have laid out a town at the mouth of the canon called "El Dorado," of which town company we are all three members. . .

If I was to consult my own interests, as certain pious and learned persons do here, I would write to you, and the rest of mankind, thus: "Pikes Peak is thirteen thousand miles above the level of anything. Eldorado is situated at its base, at the entrance to the only rich gold mines in these regions. By an easy passage up the Arkansas, you come to our town, where you can buy lots and build, and trade, or you will be furnished with long-winded, sure-footed mules, which cannot fail to take you where you can make your fortune in a day. Our town is the town. I expect the president of the

United States will come and spend a winter with us, when he gets through with his other business."

I may go to Arizona; I expect a grand rush there, from here and elsewhere, this spring. But I must finish. Yours, etc., GEO. STEPHEN.

[S. W. LEWIN] [282]
Arapahoe City, Vasques fork of Platte river,
Feb. 13, 1859.

Worthy Old Friends: – In reply to your request, I will endeavor to give you some information concerning this El Dorado of the world. I will give you a true statement. In the first place, there is gold here almost anywhere on the streams, especially those that run far in the mountains, but as for digging it out with butchers' knives, or picking it up in lumps as big as hen's eggs, I must say I have not found that place. The gold here is generally scale gold and pretty fine, and I think with proper apparatus big wages can be made.

Vasques' fork has been prospected from the mountains ten miles down stream and the prospect is from two to ten and fifteen cents to the pan. . .

No doubt you have seen samples before this; if not, I refer you to Mr. Curtis, postmaster and surveyor of the city of Arapahoe, who will carry this letter to the states with many others, being, by the way, something new to you for a man of law and order, to make the post master carry the mail.

Now for me to advise any one to come out here is more than I will do. . . We are 12 miles from headquarters now; but before mid summer, head quarters will be Arapahoe City.

---

282 *Council Bluffs Bugle*, March 23, 1859. An unsigned letter written from Auraria on February 14 was published in the *New York Tribune*, April 16, 1859. It is of the usual type.

Now my friends, facilities for writing letters here are not so great as in the states; but I will endeavor to write as often as I can; but if you miss getting a letter some times from me, don't think hard. After you read this, send it to Mr. Wicks, John Brown, Mike Buck, Mr. Ganer and so round. . . S. W. LEWIN.

[G. N. WOODWARD] [283]
Fountain City, Feb. 18, 1859.

. . . I tell you not to come here this spring, for it is just as uncertain now about how much gold there is here as it was before I left the states. I shall remain here this summer; that is, in this part of the country, to see what will turn up or down.

I have traveled this winter some distance, prospecting for gold, but have found none. So you see I have not made much.

We have had a very pleasant winter so far – no snow of any amount. Our stock have done finely, and are looking very well. We have had no trouble with the natives or Indians as yet, but expect that we may have in the spring – at least they threaten us. G. N. WOODWARD.

[J. H. MING] [284]
Auroria, K.T. February 20, 1859.

. . . I have now been a resident of this country for nearly three months and have traveled over a considerable portion of it, and being a trader here, have had an opportunity of seeing and conversing with miners and prospectors from all parts of the country. Thus I have had a good opportunity of keeping posted in regard to the country and what men are doing in

---

283 Kansas City *Journal of Commerce,* April 1, 1859.

284 *Missouri Republican,* April 18, 1859.

the mines. From my own observations, and what I can learn from others, I have at length been able to form a vague opinion, unlike many who, after a few days' residence here, tell you all about the country, its agricultural resources, as well as the auriferous richness of the gold mines. I am astonished to learn that the excitement which was created last October still continues to increase; for I have seen nothing yet which will warrant such an excitement, and when I read of the immense excitements in the states in regard to this country and look round me here, I am forcibly reminded of a certain play in Shakespeare's works [Much Ado about Nothing?], notwithstanding, there have been some discoveries made which will pay, and I believe that richer mines will speedily be opened. Each day brings to light some new development. Gold is found in small quantities in almost every place along the margin of the streams; the Platte, the Arkansas, and their numerous tributaries, but as far as thoroughly tested, but few of them prove remunerative for the labor necessary to procure the precious metal. Some miners make money, while others make none, as is the case in all gold mines. Sixteen dollars is the most that I have ever known made in one day by one man, and that in a very few instances. . .

With my experience, and from what I can learn, I do not think the average pay in the mines now open, will exceed three dollars per day per man, and perhaps much less, so you will see in my opinion that the country is much overrated in some respects, though the prospects for rich mines is very flattering. As yet no rich mines have been found.

A great many towns are being laid out in this country, some of which are building up rapidly. This, with its rival town, Denver City, hard by, seem to take the lead. They both have already several hundred good log houses, and many more, both log and framed in course of erection. They have a population of near one thousand males – two families and two or three females. Fountain City, on the Arkansas, is the next place in importance. Its population is increasing very fast with settlers from New Mexico. They are building houses, and preparing to receive the emigration. Its population is about three hundred men, three women and seventeen children. There are other towns laid out, but have not yet grown much. . .

For health, the country is not surpassed, and the climate is unobjectionable. Yours truly, JOHN H. MING.

[JOHN L. BUELL] [285]

Montana, K.T., Feb. 20, 1859.

Mr. Editor: . . . Your correspondent left Leavenworth City in company with three others, on the 20th of Sept., 1858, it being about the time when the arrival of Mr. Zimmerman in that city had somewhat allayed the excitement, a gentleman who returned from Pike's Peak, and reported, "no gold." At the time of our departure, the gold region was located at Pike's Peak. Later its location was changed to Cherry creek, and subsequently and finally to the South Platte, for a distance of about six miles on both sides of the river above the mouth of Cherry creek. We reached here in forty-six days, having come by the way of the Arkansas, and found miners making as they said, from $5 to $8 per day with a rocker. This of course we wrote home, but after a sufficient length of time had elapsed for our letters to reach their destination, we discovered that when miners rocked out this amount of gold it was done from pay dirt, that had most generally taken

[285] *Missouri Democrat,* April 2, 1859.

up the better part of two days in stripping and collecting it preparatory to washing. The weather lately has
been very mild; while I write at this P.M. the mercury
stands at 60 degrees Fahrenheit. There is nothing to
deter miners in their work if they can find diggings
that will pay. I had occasion to go down the river this
afternoon, and found just two parties at work; (six
men in all) the one was averaging $1 and the other
$1.25 per day to the man, but understood from good
authority that a gentleman by the name of Bunker,
lower down the stream, was making from $4 to $5
per day. Out of this however, he deducted one-third
of his time for stripping. . .

A prospecting party, headed by Geo. A. Thomas,
Esq., Rockport, Mo., returned last evening from a tour
into the Bayou Salada, and on the head waters of the
South Platte. Mr. Thomas, a gentleman of intelligence
whose energy as a prospector is unrivalled, and whose
judgement with respect to a gold country is unimpeach-
able, informs me "that there is no gold either in the
mountains or out of them, south or southwest from a
point twelve miles above the mouth of Cherry creek
on this river, and more especially is there none on the
head waters of the South Platte." . .

Enclosed you will find a piece of shot gold. This is
the character of the metal found on Boulder creek,
forty miles northwest from this place. It is first struck
at the mouth of the canon and found for a distance of
twenty-five miles northwesterly in the mountains, this
being as far as they have prospected.

This is the only shot gold that has been found in this
country, and the diggings are considered by a company
from Nebraska City, who have prospected them, as
far superior to anything yet discovered. Though I am

satisfied that the diggings at the mouth of the canon will pay better than those on the Platte, yet they report that for a distance of fifteen miles in the mountains the prospect increases. This will without doubt be a good trading post. The company have laid out Boulder City two miles from the mouth of the canon, which will give them the entire trade of this vast mountain region, from the fact that the pass is only accessible to mule or pony packing. The creek is about one-half the size of the Platte at this point, affording plenty of water for mining purposes, while pine in abundance is to be had for sluicing and building.

A few words as to the towns and town sites. Pueblo at the confluence of the Boiling Spring, and Arkansas river is clinging to the fond hope of gold in the South Park, El Paso to her Boiling Springs, and gold in the Bayou Salado. Montana though once a flourishing town may be considered overboard. . .

Arapahoe City, at the Table mountains, eight miles west from Auraria, is about as she was, glorying in some two or three log cabins, and several hopeful share-holders. Boulder City is in her infancy; she was laid out the tenth of this month, and already some sixty or seventy-five houses are in course of erection. The stock-holders, satisfied that they have real inducements, have set to earnestly preparing as many houses as possible to accommodate traders in the spring. Arapahoe City No. 2, or simply Arapahoe, is situated on the Cache la Poudre, at the mouth of the canon. Have not been there, but have heard that it promised well. It is my intention to make a trip to the Medicine Bow mountains the first of next month. On my return, I may possibly give you the result of my tour. Russelville, at the head of Cherry creek, and Highland, on the Platte, op-

posite Auraria, are cities which I neglected mentioning. They may be considered in a dim prospective. . .
Very respectfully, yours, etc., JNO. L. BUELL.

[A. A. BROOKFIELD] [286]
Boulder, or St. Vrain's creek, R.M.,
Sunday night, Febr. 27, 1859.

Dear Wife: – In my last letter to you I wrote that our prospects for mining were much brighter; I still feel the same. I am well satisfied that I can make at least five dollars per day. I am quite of the opinion that we have but just commenced finding what we came here for and our (Nebraska City co.) discoveries are creating much excitement at all the settlements and people are coming in every day. There are now here on the creek, over one hundred men; and we hear from numbers more that will soon be here. The gold which we find is of quite a different quality from any that has been found in other places, and what is called "shot gold" – all that has been found in other places is "scale or float gold." Before I close this I will enclose a few specimens (if I do not forget to do so). I sent you [some] sometime since, I do not at present remember where I washed it. Should any of my friends come out in the spring, give them my compliments, and tell them for me to come to Boulder creek though as I have always written I will advise no one to come; as for me I am almost certain that we are now where we can at least make good wages and I feel almost certain will make large ones. I had intended writing positively

286 *Nebraska City News,* April 9, 1859. This newspaper hails Brookfield's letter as great news and emphasizes the fact that in his previous letters Brookfield had been very conservative and pessimistic. See his letter of January 26, given above. Another letter of his, written from Boulder City, March 6, and published in the *Nebraska City News* of April 9, has a tone of hopefulness.

for you to come on, but on reconsideration I think it will be much pleasanter that you (if you come) start later in the season, as then the roads will be much better, and also the weather. A. A. BROOKFIELD.

[UNSIGNED LETTER] [287]

Fountain City, March 1, 1859.

. . . Fountain City is situated at the confluence of the Arkansas river and Fountaine Qui Buille; it contains about forty houses, and has the appearance of quite a town. The country surrounding it is well adapted for agriculture, and in this respect has the advantage over other towns which have sprung up within a year on the Platte and Cherry creek. . .

Mr. Ming, by whom I send this letter, informs me that he has come from the Platte, and that people there have found better diggings than ever, and everybody seems to be encouraged; they make now from $10 to $20 per day. I have not been mining yet, nor do I intend to follow that branch of business. I have been prospecting about a little, but did not find enough of the dust to encourage me to operate any further unless I can make it pay largely. I am now engaged in surveying, and many other things that pay me better than digging. S— and myself have formed a partnership as surveyors, and have more work on hand than we can accomplish in three months hence. We make as high as $60 per day in surveying town sites and claims, but the great difficulty is that we cannot get but very little cash for our work at present. We have to take stock and orders on stores for pay, but expect there will be plenty of gold and money in the spring. The large body of gold is supposed to be on the head waters

[287] *Missouri Democrat,* April 25, 1859.

of the Arkansas, but it is too cold now to go in the mountains to prospect. . .

[E. E. KIRK] [288]

Denver City, K.T., March 1st, 1859.

Mr. Thomas A. Kirk: Dear Brother – I have had my health better out here than I have had for the last fifteen years.

I have been searching for gold all winter, and I have found it at last. One of my crowd has just returned from out of the mountains and found a place where he can take three hundred colors to the pan.

Get one or two good mules and pack through. Do not wait to join any company; one or two men are as many as I would want.

Provisions are as cheap here as you can buy them and haul them.

Do not pay any attention to men coming back disheartened, for they are too lazy to work good mines, when they find them.

You need not bring any picks or shovels as I have plenty of them.

Direct your letters to Denver City, K.T., via Ft. Laramie. Your most affectionate brother, E. E. KIRK.

[JOHN SCUDDER] [289]

Auraria, K.T., March 4, 1859.

The express leaves in the morning for Fort Laramie, and I will improve the opportunity to give you some news from this far distant land. The most important

---

[288] Published in the *St. Joseph Gazette;* the *Hannibal Messenger,* April 10, 1859; and the *Nebraska Advertiser,* April 14, 1859. A letter written by A. J. Pullman from Denver City, March 1, was published in the *Leavenworth Times* and copied in the *Kansas Tribune* of April 14, 1859. It contains no important news.

[289] *Missouri Republican,* April 13, 1859.

to your readers will be the gold news, so I will give that first. All the statements made in my last letter, which reached you in the February mail, have turned out to be true, so far, and many more that I could tell you about, if I could remember them. About forty miles directly north of this place, at the base of the mountains, on what is known as St. Vrain's creek, a party of men have discovered the shot, or nugget gold. They came down here to purchase tools and provisions, and spent some two or three hundred dollars, and paid their bills in that kind of gold. The largest lump that I saw from those diggings was worth two dollars and sixty cents, and from that down to ten cents. Two miles from this place a party of men, six in number, have opened some diggings on the Platte, about two hundred yards from the river. One man works a rude pump, which forces the water through a cotton hose made from wagon covers, old clothes, etc., while the others carry the dirt in buckets and on shovels to the rockers to wash it. They have worked in this place two days. They took out the first day thirty dollars, and the next forty dollars. I rode up to see them this evening, and the result of their labor, they could not tell, but think this day's work will turn out better than the other two. Out of the thousand or twelve hundred men in the country, I do not think that fifty men are at work mining. The weather is too cold for some, and the most of them are engaged building houses in the various town sites. During the last week we have had about two hundred arrivals from New Mexico, and among them a good number of ladies. I returned to this place from Fort Laramie last week, where I have been to solicit the aid and protection of the commanding officer at that post from the Indians. They have not troubled us much this winter, but we are entirely at their mercy,

and during this month and April we look for them
to take as much of our stock as they want.

The Arkansas from Bent's Fort up to the mountains
is one vast Indian camp – the Camanches, Kioways
and Arrapahoes are all banded together, and unless the
government does something to protect the emigrants,
many, I am afraid, will never reach this place, or
return to their homes in the states. There is a movement
on foot to petition the government to have the post
now located at the Chinese pass,[290] which is of no use
whatever, where it is, to this place or somewhere on
the South Platte. This change has been long talked
about, and is now of more importance than ever. The
best road from the states to the mines is to embark at
Kansas, or any point between that place and St. Joseph,
go direct to Fort Kearny, then up the South Platte to
this place, which is the depot or outfitting post for the
whole country. This place has been built as if with
magic hands. The houses number over four hundred,
with about six hundred regular inhabitants. We have
six good trading houses, two blacksmith forges in full
blast, and grog shops without number. The gamblers
are very numerous, and get most of the gold in their
possession. They have come in from Salt Lake, New
Mexico and the various forts through the country
until the place is full of them. They brought money
and horses with them, and many of them are quite rich
for men in this country. . .

The receipts of a post-office here next season will be
as heavy as that of St. Louis, and perhaps more. Our
mail last week weighed 400 pounds, and mostly all
letters at that. . .

---

290 Shiann (Cheyenne) pass (Camp Walback), twenty miles east of present
Laramie, Wyoming. See Coutant, *History of Wyoming*, 368.

We have organized a claim club – a sort of copy of the United States preemption law – and about two hundred men have made, or intend to make, farms. About sixty per cent of the county is of no value for farming purposes, but will make excellent stock or grass lands. . . Truly yours, JOHN SCUDDER.

[JOHN POUDER] [291]

Auraria, K.T., March 5, 1859.

Pete: – I received your welcome letter, and was somewhat glad to hear from you. The weather is very fine, and it looks like spring, and everything is going on well – gold is getting to be more plenty – men have got out of their shanties, and gone to digging. . . Three miles above here there are several at work. [Henry] Allen, the postmaster, is at work there. Here (enclosed) is ten dollars worth of the precious metal, which was dug out by him in one day; and he says there is a man there working a claim which pays 25 cents to the bucket. Allen is positive that when they get their sluices to work, six or eight men can make upwards of $100 a day. . .

There is a company going to start for the "Parks" [South Park and Middle Park?] in a few days. I think there is something rich in them, for some men started a new town five miles above Russelville, and got out logs to build; in the meantime part of them went through to the "Parks," they came back, sold out, and all hands packed up and left for there again. It is reasonable that they found something that must pay very well, for they left a place which paid a man well – some $3 to $10 per day with a pan and rocker.

Pete, I would like to have you come out as soon as

---

[291] *Missouri Democrat,* April 16, 1859.

possible – bring out two light spring wagons, four
horses and harness and tools. Your partner and friend,
JOHN POUDER.

[A. A. BROOKFIELD] [292]
Boulder City, Rocky mountains, N.T. March 5th, 1859.
    Friend [Levi P.] Morton: [tells of shot gold, etc.]
. . . We have found this gold in nearly every place
we have prospected from the mouth of the canon, for
a distance of twenty-five miles in a north-westerly direction, and I know, by actual experiment, that a man
can make with a rocker $5 per day. . .

    I did not come out here for town speculation, but
after we made what are considered by far the best
discoveries of the "precious metal," we thought as the
weather would not permit us to mine, we would lay
out and commence building what may be an important
town [Boulder], provided the mines prove of the richness which we expect. – this, however, is only a second-
ary consideration. . . Yours truly, A. A. BROOKFIELD.

[WILLIAM LARIMER] [293]
Denver City, K.T., March 19, 1859.
Editor Times: – It is now over a month since I wrote
to you. That short month has been well improved by
our people, in the way of building houses, prospecting,
etc., and as springtime is now on hand, here we are all
excitement in relation to the future of this beautiful
country. We hear every day of the vast emigration that
is coming here. Over one hundred wagons are now

292 *Nebraska City News,* April 16, 1859.

293 *Leavenworth Times,* April 23, 1859. John Scudder wrote to the "president of the chamber of commerce of St. Louis" from Auraria on March 20
a letter that was published in the *Missouri Republican* of May 12, 1859. He
spoke of the great resources of the new mining country and advised the
buying of supplies at St. Louis.

this side of Fort Kearny, on their way here, with one steam saw mill from Harrison county, Iowa. . .

Only a part of our emigration is expected from the east. Within the last week twenty-five heavily loaded teams, with over fifty Mexicans, have arrived from New Mexico, and they are coming in daily laded with all kinds of Mexican productions, such as flour, bacon, beans, potatoes, onions, etc.

I have written you often about the chances of this becoming a farming country, and I am more and more confirmed in this opinion every day. I am told by the Mexicans themselves that our land is quite equal to theirs in every respect, . . .

We have now four stores in Denver City and two in Auraria, all belonging to one merchant from New Mexico, except one firm, Blake & Williams, from Council Bluffs, Iowa. The people here want for nothing; still provisions are very high. Flour is selling at $15 per 100 lbs; bacon, 50 cts per lb; sugar, 50 cts. Building material of all kinds are wanted badly. I have heard of 75 cents a pound for nails, and none here. Nails is the constant inquiry. Glass and paints, in fact everything in the building line is wanted.

This country is shaping matters for a glorious future by establishing all kinds of societies, such as, historical, pioneer, etc., also, laying off lots for all the different denominations to build churches and schoolhouses; also, beautiful cemetery grounds, called Mount Prospect. I attended the first funeral to-day, at the new cemetery, and the first in Denver City. A Mr. Edward Hay, of De Soto, Nebraska, died up in the mountains, about twelve miles from here, on the 17th. The deceased was rather a young man, very healthy and robust. He had been suffering from a severe cold, and went to the mountains to hunt. He had been very successful, and got over-heated, and took a sudden cold from the mountain winds, and died suddenly, with swollen glands, in a few hours. There had been two deaths, previously, up in Montania; one of them, young Boucher, of Oskaloosa. He came out in our train. The other was a young man – Henry Britingham – formerly from St. Louis, and recently from Santa Fe. These three deaths are all that have occured in this country, that I have heard of. . .

Messrs. A. [St.] Vrain & Co. have established a very large store of provisions and merchandise in Denver City. Mr. Edmond A. [St.] Vrain is nephew to Col. Ceran St. Vrain, of St. Vrain's Fort. The firm is putting up a very large warehouse, 30 by 80 feet long. Our buildings are nearly all hewed pine logs.

Our people are all in high spirits in relation to the prospects of this country, with an occasional exception. . . Yours truly, WM. LARIMER, JR.

THE GOLD RUSH OF 1859

THE GOLD RUSH OF 1859

EMIGRANTS TO THE GOLD REGIONS [294]

We were yesterday in conversation, at the Union hotel, with Mr. J. P. Hecock, of Pennsylvania; Mr. J. Cairey, of Illinois; and Mr. T. Benjamin, of Iowa. These gentlemen make three of thirteen men, who are on their way to Pike's Peak. They speak in the most extravagant terms concerning the emigration that will set towards the Rocky mountains in the spring.

Mr. Hecock says that New York and Pennsylvania will send at least twenty thousand people, mostly families.

Mr. Benjamin, of Iowa, says that every family in that state will send one or more representatives.

Mr. Cairey, of Illinois, informs us that he cannot be more definite than to say that the whole state is coming.

They further inform us that over one thousand people from Illinois and Ohio are now on their way to this city. . .

GOLD MINES [295]

A person who comes from the gold diggings, without being able to pronounce them equal to California, must either keep it to himself, or expect to be bully-ragged soundly, and made out to be a perfect ignoramus, not knowing what gold is, because he cannot see

[294] Kansas City *Journal of Commerce*, January 16, 1859.

[295] The *Kansas Chief*, January 20, 1859. Mr. Solomon Miller, publisher of the *Kansas Chief* of White Cloud, was one of the few editors who was very skeptical of the stories of gold discoveries. Throughout the winter and spring of 1859 he continually warned against over-optimism and published the un-

it lying about in piles, where it is not. This is the course pursued towards Henry Sterett, of Holt county, Mo., who recently returned from Cherry creek, and honestly gave his opinion of it, which was published. We have seen articles in at least two papers, not very complimentary to him. In plain terms, they make him appear an ignoramus, because he saw proper to tell the truth. It is admitted that he has been to California, where he was engaged in mining; yet he knows nothing about gold digging, because he does not help keep up the excitement; while persons who have never seen gold dust, and have rarely possessed the metal in coin, are quoted as reliable testimony, because they bring extravagant stories. If there is gold at Pike's Peak and Cherry creek, the fact will be fully established, without this treatment of persons who have failed to discover it, and honestly say so.

PIKE'S PEAK [296]

Ho for Pike's Peak! There is soon to be an immense migration, especially from our western states, to the new El Dorado. The extensive failure of crops in 1858, the universal pressure of debt, the low prices realized or promised for the fruits of the husbandman's labors, the deadness of enterprise, the absence of thrift, render such migration inevitable. There is scarcely a village west of Ohio in which some are not fitting for and impatiently waiting the day when a start may be prudently made for the neighborhood of Pike's Peak. We shall

favorable reports. In his issue of February 17, 1859, he says: "There has been considerable grumbling at us, on account of the course we have seen proper to pursue in regard to the gold mines. Now, our belief in the matter, does not need to influence any one else. If persons wish to go to the mines, let them go—we cannot prevent them, and they have no need to sit down and grumble at us for it. Our disbelief in the abundance of gold, will not change the reality one iota. . ."

[296] Editorial in the *New York Tribune*, January 29, 1859.

be disappointed if less than fifty thousand persons start for the new gold diggings within the current year. Many go to dig, perhaps quite as many to speculate on the presumed necessities, or fancies, or vices, of the diggers. Liquors, tobacco, etc., will be dispatched in quantities. If the Pacific Railroad were this day completed on the central route, we are confident that the eastern third of it would pay a fair interest on its needful cost during the year 1859 . . .

[GOLD RUSH ARMY][297]
. . . A bigger army than Napoleon conquered half of Europe with, is already equipping itself for its western march to despoil the plains of their gold. The vanguard has already passed the Rubicon, if I may so metamorphose the muddy Missouri. . .

CARTS FOR GOLD HUNTERS [298]
There was yesterday upon our streets a novel sight – something that the oldest travelers over the plains had never seen. It was a cart drawn by an ox in harness. There is not a piece of iron of any description about the cart. It is made upon the same plan as those of the Red river people, a very minute description of which is in the January number of *Harper's Magazine*. The whole cost of the cart, ox and harness was $58, and with it a man can carry from ten to fifteen hundred pounds, all depending upon the vim of the ox. These carts are certainly the best contrivances for gold hunters to move over the plains with speed and certainty, that we have

---

[297] Correspondence from Des Moines, Iowa, dated January 31, 1859, and published in the *Chicago Press and Tribune*, February 4, 1859. The correspondent tells of the return of Valorious Young of the Russell prospecting party, relates his experiences in the mines, and says that he is "preparing to go to the diggings in the spring, taking with him a saw mill (Query—who isn't going to take a saw mill?)."

[298] Kansas City *Journal of Commerce*, February 1, 1859.

ever seen, and we feel quite confident that they will be used by the emigration in large numbers. The cart on exhibition was made at Westport.

OFF FOR THE MINES [299]

A German, solitary and alone, all the way from St. Paul, Minnesota, on foot, passed through our city on Thursday last, bound for the gold region. His out-fit consisted of a rifle, a large butcher knife, and the clothes he wore. On being asked if he expected to reach the gold mines on foot he replied: "Yaw; me has walked vrom St. Baul, he furder nor dat."

A HAND CART TRAIN [300]

A company of one hundred men has been nearly raised in this city, for the purpose of starting to mines with hand carts – a la Mormon – between the fifteenth and twenty-fifth of this month. Each hand cart will be manned by from three to four men, and be freighted with one hundred pounds to the man. The whole cost of their outfit – carts, clothing and provisions – the Omaha market, will be about twenty dollars per man. This train will go through quicker than mules. One hundred pounds each, will enable them to take provisions enough to last them through, with two week's supply after they get there; three month's clothing; and all the necessary cooking and mining utensils. The labor of pushing their carts through will be nothing; those who go out with teams have to walk, the hand cart men are even with them there; those who take teams have to

---

299 *Council Bluffs Bugle*, February 2, 1859.

300 *Omaha Times*, February 3, 1859. The same paper tells that a train of fifty wagons will leave between the 10th and 20th of February for the mines, as "the advance guard of the Omaha delegation." W. N. Byers, pioneer newspaperman, is listed as one of these.

herd their cattle and mules nights, and charge around
for an hour every morning hitching up, the cart men
can lay down to sleep, and in the morning take the road
without delay; teamsters have to depend on grass, cart-
men can camp any place; teamsters have their teams
and provisions to take care of when they get to the
mines, carts will be worth ten times their original cost;
wagons will be useless, carts a convenience. We feel
proud of our hand cart boys, and every person who be-
comes one of their number and goes through, shows
himself a man – undaunted by danger or fatigue and
undiscouraged by difficulties. . .

[THE FIRST TOPEKA TRAIN][301]

The first Topeka Pike's Peak company will leave
here for the mines, about the middle of the present
month. This, however, is a "packing" company – the
one intending to use wagons will probably not leave
before the first of April.

FOR PIKE'S PEAK [302]

Parties are beginning to move westward to the new
gold fields. We hear of two companies who went west
within the last week from Clinton, Iowa, by the Chicago, Iowa and Nebraska Railroad, and of another that
took the same direction by teams. Within the next few
weeks we presume all our thorofares west will be
crowded with emigrants.

THE VERY LATEST FROM THE MINES [303]

We had yesterday the pleasure of meeting Messrs.

301 *Kansas Tribune,* February 3, 1859.
302 *Chicago Press and Tribune,* February 4, 1859.
303 *Nebraska Advertiser,* February 4, and copied in the *Missouri Republican,* February 11, 1859.

C. A. Lawrence, Dr. A. S. Kunkle, J. W. Winner and
James Hall, who arrived at this place directly from the
gold regions – Denver City. They made the trip in
twenty days, . . .

From these gentlemen we gain the most valuable and
reliable information. Mr. Lawrence is of the "Lawrence company," that left Leavenworth last fall; Dr.
Kunkle is from Sioux City, Iowa, and went out by the
Omaha route; Mr. Winner is from St. Louis – son of
ex-Mayor Winner – and Mr. Hall is from Leavenworth.

These gentlemen have thoroughly prospected, and
exhibited specimens of the precious metal. Mr. Lawrence is an old California adventurer, and has no hesi-
tancy in pronouncing the Cherry creek mines equal, if
not superior, to any of those of California. Those now
mining, under the many disadvantages, are making
from $2 to $12, and often $15 per day. . .

Gold has been discovered in satisfactory quantities,
along the base of the mountains, from New Mexico to
Fort Laramie. . .

It is needless to say that this encouraging and reliable
news has given new impetus to the gold excitement in
this vicinity. Many who were before disposed to
"doubt," no longer "hesitate," and will be off in the
"first boat."

PIKE'S PEAK GOLD AUGER [304]

There are many things gotten up merely to sell; but
a little the best article we have seen for actual use is a
"prospecting auger," manufactured at Hood & Langan's shop on Locust street. By the simple act of boring,
in 15 minutes the soil ten feet below the surface can be
brought up, washed out, and its gold character ascer-

[304] *Missouri Democrat*, February 8, 1859.

tained. With this auger miners can prospect with little labor all the way to the mines, and select the best at once. The same Hood & Langan beat the world at boring; they are fully up to the times, and made the pick that struck the first lick in the Pike's Peak gold mines of Kansas.

A NOVEL MODE OF TRAVELING [305]

Two gentlemen by the name of I. J. Stevens, and N. Smith, arrived in this city a few days since, from Minneapolis, Minnesota, on foot, a distance of 460 miles, and left this city on Thursday, for the gold mines. They have an entirely new mode of traveling – far eclipsing the hand cart or wheel borrow – it is called the "Tebogia," a novelty in this section, but are common as a mode of conveyance between the Selkirk settlement and St. Paul. Its construction is very simple and light, being a piece of sheet-iron 8 feet long, 18 inches wide, turned up slightly in front, to allow it to pass over low obstacles readily, and 4 inches deep. To this, the two men are attached, by harness. In this, they take out sufficient provision and clothing to last them through their long, tedious and dreary journey, across the plains to the gold mines. They are tough, hardy looking fellows, and can endure the many hardships and privations experienced in a trip of this kind, in the dead of winter. We like to see such men succeed, and hope they may be abundantly paid for their trouble and hardships.

FOR THE GOLD MINES [306]

A party of three men, from Morris, Grundy county, Ills., passed through this city yesterday, on their way

---

[305] *Council Bluffs Bugle*, February 9, 1859.
[306] *Omaha Times*, February 10, 1859.

to the gold mines. They came through on foot, having thus traveled nearly as far as the distance from here to the mines, and intend going through to Cherry creek in the same way, having obtained a hand cart at Council Bluffs, in which they placed their outfit and provisions enough for the trip. They report that the excitement is very great in Illinois, and that "every other man is going," to use their own words. They also state that a company of about 400, which is formed at Morris, are having a steam wagon of great power constructed, with the intention of trying the experiment of traveling to the mines with it.

VERY LATE FROM THE MINES [307]

Yesterday evening Mr. E. A. Muir, of the firm of Goddin, Miller & Co., of this city, returned home, direct from the gold fields of Nebraska. There was an intense interest throughout the city to learn the latest news from the mines. . .

Mr. Muir states there are, at present, about 700 men in the mines, all healthy and in fine spirits, elated with present and future prospects. . . Mr. Muir states that gold exists throughout the country; that it is difficult to find a shovelful of dirt that does not contain more or less of the precious stuff. . .

The Nebraska City boys have found, and were the first to find, shot gold. They are consequently in exuberant spirits. They cannot be hired to return. . .

Mr. Muir made the trip in the depth of winter with two yoke of oxen, and Mr. Stufft, of Otoe City, as his only companion. . .

[307] *Nebraska City News,* February 12, copied in the *Missouri Republican,* February 20, 1859. The *St. Joseph Gazette* of February 14, 1859 records the return of J. S. Tutt from the mines with a favorable report.

EMIGRATION – FIRST INSTALLMENT OF THE SPRING [308]

A company of twenty-four persons arrived in our city last night, en route for Pike's Peak. A gentleman who came in company with them was in our office just before we went to press, and informed us that they were from different states of the north and east, and all make one report that the emigration will be immense, probably reaching one hundred thousand. Most of the company will remain in the city until a suitable time for starting, a few of them, however, going into the country, on account of the cheapness of living.

THE RUSH FOR PIKE'S PEAK "GOLD MINES" [309]

The glowing accounts of the richness of the alleged gold region of western Kansas and Nebraska, are having the effect that the speculators at St. Louis, and on the Kansas and Nebraska borders, who write them, designed they should have. There is no doubt but that more than one-half that comes to us, in the shape of letters and through newspapers, regarding these gold discoveries, are monstrous exaggerations, and, perhaps, totally false and baseless.

However, those who choose, may "try their luck" at Pike's Peak, in utter disregard of the caution we suggest. We advise no man to go there until the thing is "certain sure," which now it is not, by a great deal.

We have within a few days taken pains to examine our western exchange papers with a view of ascertaining the extent of the probable emigration to the reported gold regions. The result of our examination may be found below.

The *Peoria Transcript* says a company has been or-

---

[308] Kansas City *Journal of Commerce,* February 13, 1859.
[309] *Missouri Republican,* February 16, 1859.

ganized in the eastern part of this state with the design of going to Pike's Peak in the spring.

The *Dubuque Times* says "numerous families west of Dubuque are making preparations to go to Pike's Peak in the spring."

The Sycamore (De Kalb county, Ill.,) *True Republican* says: "An organization has been perfected by the Pike's Peak enthusiasts of our village, and a president, secretary and treasurer elected, and they hold evening meetings at the American House."

The *Woodstock* (McHenry county, Ill.,) *Sentinel* is informed that "a number of our citizens are preparing to take a trip to Pike's Peak."

The *Detroit Tribune* learns that a number of young men in that city have formed themselves into a company to go to Pike's Peak.

The *Columbus* (Ohio) *Gazette* says not less than a hundred young men will leave that vicinity for the reported gold mines in the spring.

The Joliet (Ill.) *True Democrat* says a number of young men are forming Pike's Peak companies in that place.

A party from Lockport are making preparations to start soon.

The Davenport (Iowa) *Democrat* says three men with their wives and three children, passed through that place in wagons the other day, en route for Pike's Peak.

The Bloomington (Ill.) *Pantagraph* says, some twenty-five young and middle aged men in that city, will leave for Pike's Peak and Cherry creek before the middle of April.

The *Amboy* (Ill.) *Times* states, that a company has been organized for Pike's Peak at that place, and among its members are the mayor and other citizens of high standing. The same paper says other companies are forming there, and at Lee Centre. The first named company intend to start in ox teams the present month.

A company has already set out from Dixon, Ill., for the mines.

At Iowa City, some fifty will leave for the reported gold region in the spring.

One company of twenty-five, and another of about one hundred, under the auspices of what is called the Indiana and Illinois Pike's Peak Mining Company, will start within a few weeks from northern Indiana and the Illinois counties adjoining.

A company passed through Columbus (Ohio) for Pike's Peak, a few days ago, designing to go right through.

We have seen other accounts from various localities, where efforts are in progress to organize companies for the same purpose. There are doubtless many other localities we have not heard from, where this gold excitement runs equally high, and where emigration companies are forming, or are already organized.

The *Iowa City Republican* says a party of three young men from Chicago passed through that place a few days since, en route for the gold regions; and also that Mr. B. [R.] Sopris [310] of Michigan City, acting on behalf of the Indiana and Illinois Pike's Peak Mining Company, is expected there soon with two companies for the same destination, one of twenty-five men, and the other of from seventy-five to one hundred. The railroad from this city to Iowa City, says the *Republican,* has agreed to reduce the fare $2 per man for these two companies, and the Western Stage Company have agreed to take them from Iowa

[310] Captain Richard Sopris and his family became prominent pioneers of Colorado.

City to Council Bluffs at a reduction of $3 per man. The same paper says a company of six from Dixon, Ill., passed through that city, "armed and equipped" for the mines, a day or two since. The *Republican* says, fifty residents of Iowa City intend starting for Pike's Peak during the spring.

[A COMPANY FROM GEORGIA] [311]

A number of young men of Lumpkin and the adjoining counties are making preparations to start to the lately discovered gold mines of the Rocky mountains. This company is being conducted by G. M. [W. G.] Russell, of Dawson county, who has recently returned from that section of the country, and who is an experienced miner of the first water. It is the purpose of the company to start about the 15th of March next.

[LONG TOMS AND ROCKERS] [312]

The undersigned are manufacturing long toms, rockers and all kinds of mining implements suitable for the Kansas mines, at No. 5 Washington avenue, between Main street and the Levee, and will keep a supply constantly on hand. The long toms are so made that they only weigh from thirty to forty pounds; rockers from twelve to fifteen pounds. Mr. Carman having eight years experience in California, will give any information in regard to mining. Persons outfitting for the mines will do well to call and see them. The expense of taking long toms from this city to Pike's Peak is about $4; rockers $1.25. CARMAN & McCULLEN.

[311] *Macon* (Georgia) *Journal and Messenger*, February 16, 1859.
[312] Advertisement in *Missouri Republican*, February 16, 1859.

N.B. Persons sending orders from the country will be promptly attended to.

[THOSE GOING TO THE MINES] [313]

. . . As usual, those who are doing well now, pecuniarily, are the uneasy men who are anxious to do better. . . (Those who are not doing well seem not to aspire to anything better.)

Providence blesses with the great solace of contentment those to whom she denies the gifts of fortune. This will be found to be the general fact. The man who prospers above his fellows is always the first to seek new enterprises in the hope of gaining money still faster, and those who are now talking most earnestly of a venture to the gold diggings are those who have good situations and good pay at home.

WHO SHOULD GO TO PIKE'S PEAK [314]

We often hear young men, who never did any hard work in their lives, talk about going to Pike's Peak. We ask such what kind of work they think gold digging is? Let them turn out here and get themselves into practice by digging wells, cellars, coal, quarrying rocks, mauling rails and rolling saw-logs, and eat dry bread and wash it down with water, and sleep on the ground in fair weather and foul, and then form an opinion about the work of digging gold. Digging gold is no child's play; and it is only the strong, able-bodied, hardworking men that will suit for the business. The men who succeeded at the mines in California, were strong of arm, stout of heart, and only such men can

[313] The *Springfield* (Massachusetts) *Republican,* reprinted in the *Chicago Press and Tribune,* February 17, 1859.

[314] The *Leavenworth Ledger* of February 17, 1859; quoted in the *Missouri Republican* of March 12, 1859.

succeed at the gold mines. A bull has just as much business in a china shop as a glove-handed clerk or fair-faced mechanic has at Pike's Peak.

[EQUIPMENT FOR THE MINES] [315]

The following list embraces what is actually needed by those going to the mines. The list of course can be extended in accordance with desires and the purse. The cost of these articles will be about $150.00 to the man, or $600.00 for the company, and will by careful and frugal management, last them for six months. . .

[There follows a list of supplies. This is similar to lists given in the preceding volume (IX) of this series.]

MASS MEETING AT KANSAS CITY [316]

Judge B. A. James was called to the chair, and C. C. Spalding was appointed secretary.

Mr. Groom was solicited to state to the meeting the object for which this mass meeting had been called.

In an elaborate and eloquent address, he stated that Kansas City was far behind other points upon this river in their efforts to secure the business that this great emigration will engender. He showed to our people the influences that Leavenworth and St. Joseph had set to work to direct this emigration and its business to their respective cities.

Already Leavenworth had succeeded in directing the first emigrants to that point, and unless we go to work in earnest, all the trade and traffic which will be engendered by this great exodus, will go by us and concentrate itself at Leavenworth. . . Because God in his providence, has given us superior advantages

[315] *Nebraska Advertiser,* February 17, 1859.
[316] Kansas City *Journal of Commerce,* February 18, 1859.

over any other town upon the Missouri river, this is
no reason why we should remain idle.

All this can be overcome by the influence of capital.
Capital can accomplish anything – it can control politics, religion and commerce, and unless we bring to
bear the agencies that capital can accomplish, we certainly shall lose most, if not all the advantages that
will be engendered by this immense emigration.

On motion of Col. E. C. McCarty, a committee of
five was appointed to report business for the meeting.

EMIGRATION [317]

We see from our exchanges that the hotels of St.
Louis are filled with people waiting to go up the Missouri river. Hundreds have taken quarters aboard the
boats, paying $.50 per day for their board. They are
bound for the gold mines.

EXPRESS TO THE MINES [318]

G. W. Smith's package and passenger express, left
this city yesterday for the gold regions, via the great
Santa Fe route. This is the first express of the season,
and Mr. Smith informs us that he expects to make the
trip in fourteen days at the outside.

CATTLE FOR THE GOLD HUNTERS [319]

We learn from Col. Nelson, that Messrs. Coleman
& Turner, great cattle drovers, have now upon the road
fifteen hundred head of cattle that they are driving
from western Texas to this market for the express purpose of supplying the emigration of gold hunters. These

317 Kansas City *Journal of Commerce*, February 20, 1859.
318 *Ibid.*, February 22, 1859.
319 *Ibid.*, February 22, 1859.

cattle are all large and in fine condition, and selected expressly for this market. They are all well broke, four years old and upwards. They will be here at an early day. Emigrants will please take notice.

ITEMS ABOUT THE MINES [320]

—The Pike's Peak excitement was running mountain high in this city yesterday. About a hundred gold seekers came in on the trains last night. The hotels are becoming crowded from dome to cellar. – *St. Joseph Journal,* February 18th.

—Jack Henderson left St. Joseph, with a party of fellows on the 17th inst., for Pike's Peak.

—A steam mill, drawn by eight yoke of cattle, passed through Omaha, N.T., a couple of weeks ago, bound for the mines.

—Great numbers of the Utah saints are putting their households and harems in order, preparatory to leaving for the Pike's Peak mines in the spring.

—The company organized in Washington city to run a daily line of stages to Pike's Peak from Leavenworth, has a capital of two hundred thousand dollars. W. H. Russell, of the firm of Major & Russell, is president of the company. . .

FOR THE MINES [321]

A company of young men under the command of Capt. Henry Judd, leave here on the 10th of March, for the Peak – they pack their outfit on ponies. The first cattle train will leave on the 10th of April. Anderson & Co. propose to carry out 20 passengers, allowing 100 lbs. of baggage to each man, board them on the way out and one week after their arrival there, for the

---

[320] *Lawrence Republican,* February 24, 1859.
[321] *Kansas Tribune* (Topeka), February 24, 1859.

sum of one hundred dollars. If you want a passage, apply soon to G. W. Anderson, Topeka.

FIRST COMPANY FOR THE GOLD FIELDS [322]

The first company for the Kansas gold mines leaves this city next Monday, under the charge of Capt. A. Cutler, our present most worthy and efficient city engineer. The company is composed mostly of enterprising young men, and will of course take the great central or Smoky Hill route to the mines. They take up the Kaw river as far as Salina, and thence almost on a straight line up the Smoky Hill to El Paso,[323] immediately at the base of Pike's Peak, and the county seat of the newly erected county of El Paso. The distance by this route from the city is only 470 miles – or over 200 miles shorter than by any other route, besides finding wood, water and grass in much greater abundance than by either the northern or southern routes. The

---

[322] *Lawrence Republican,* February 24, 1859.

[323] The *Lawrence Republican* of March 3, 1859 says:

"Capt. A. Cutler left for the mines on Monday last, with his company. He takes the Smoky Hill fork. His business is at once to complete the survey of El Paso, the flourishing county seat of the recently erected county of El Paso. This is a town recently laid out at the foot of the gold mountains. It is the termination of all the routes to the mines, and is the point where wagons must be discontinued and pack mules employed to go into the mountains— a kind of second Sacramento. From the map and description of the country, we see that it is a beautiful region, and abounds in precious pine and cottonwood. Mr. Cutler, with other gentlemen, purpose making large investments there. Success to them."

In the *Herald of Freedom* of March 19, 1859, we read: "The El Paso Town Company have made arrangements with S. S. Prouty, late of the Prairie City *Champion,* to establish a paper in that place. The press and material will go forward about the first of April. El Paso is at the foot of Pike's Peak, and it is said, is destined to be the city of the gold region. It is represented that shot gold has been found in the immediate vicinity of the town."

In the same issue of the paper the editor says he was shown a beautiful map of El Paso, representing the town and the newly-organized counties of El Paso, Oro, etc. The *Leavenworth Times* of March 19, 1859 says that S. S. Prouty is to publish the *Champion* at El Paso.

company got their entire outfit in Lawrence. They send their teams with provisions as far as Salina, and from thence will travel with pack Indian ponies, and expect to make the gold regions in fifteen or twenty days from Salina. . .

FOR PIKE'S PEAK BY WATER [324]

Capt. J. Greenhalgh, says the Cleveland (Ohio) *Leader,* owner of the steamer tug, Niagara, proposes to take passengers to Pike's Peak via water. Should a sufficient number of passengers offer, he will make the voyage through the Ohio canal to Portsmouth, down the Ohio to the Mississippi, up the Mississippi to the Missouri, up the Missouri to Kansas City, up Kansas river to Fort Riley, and from there up the Smoky Hill fork as far as practicable towards the mines. This will be a voyage of nearly or quite two-thousand miles, and we doubt very much whether the trip comes off.

PIKE'S PEAK A HUMBUG [325]

A man calling himself Mr. Ross, appointed Friday evening, in Davenport, to deliver a lecture on Pike's Peak. The *News* says:

"At the hour appointed the people commenced flocking in to hear all about the gold mines at Pike's Peak, from one who had been there. A man with a false moustache and hat down over his eyes, took in the quarters at the door. How many he took in is not known. At any rate he took in the audience. For, after waiting patiently for Mr. Ross and his eloquence, for some time, some inquisitive persons went out to see the cause of

---

[324] *Leavenworth Times,* February 26, 1859.
[325] *La Grange* (Missouri) *National American,* February 26, 1859.

the delay. When lo! the doorkeeper, having made his haul, had 'scooted,' and was no where to be found." Ross must have convinced his audience that either himself or Pike's Peak was a humbug.

INQUIRIES ABOUT GOLD [326]

We receive letters by scores about the gold mines. The old and young, bald and beardless, masculine and feminine propound innumerable inquiries and request immediate replies. Friends, it is a sheer impossibility. We publish all the intelligence which reaches us concerning the mines, and those who seek information should send for *The Times* rather than write private letters to the editor.

CHEAP AND EXPEDITIOUS PASSAGE TO PIKE'S PEAK [327]

The undersigned would inform those who contemplate a trip to the great El Dorado of the west, that he has completed arrangements by which EDWARD W. WYNKOOP, ESQ.,[328] sheriff of Arappahoe [*sic*] county, has consented to conduct a train from Lecompton to Pike's Peak.

Sheriff WYNKOOP has but recently returned from Pike's Peak on business connected with that district. He has taken notes of all matters of interest of importance to emigrants along the different routes, and

[326] *Leavenworth Times,* February 26, 1859. In the same paper, A. J. Bowen advertises as a guide for gold seekers, saying he had been to the mines the previous summer. He advises against starting until grass makes its appearance.

[327] *Herald of Freedom,* February 26, 1859.

[328] Wynkoop and A. B. Steinberger left Auraria on December 3, 1858, and arrived at Omaha on January 5. The *Omaha Times* of January 6, 1859 gives the report of their mid-winter trip. Wynkoop froze his feet, but not severely. After his return to Colorado, Wynkoop enlisted in the military and was a major during the Civil war. One of the first streets of Denver was named for him.

is acquainted with the most favorable points for water-
ing, wooding, grazing and camping. This knowledge
is all-important to those journeying thither, and the
present must be regarded by all as a most favorable
opportunity to traverse this otherwise almost impene-
trable country.

From Lecompton, Kansas. Between the 1st and 15th
of April, 1859. Persons who may see proper to join our
party, will be provisioned from the date of starting
until the expiration of one week after arriving at
Denver City, (which is located at the mouth of Cherry
creek, in Arappahoe [sic] county,) on the following
terms:

One person, with one hundred pounds of baggage,
ONE HUNDRED DOLLARS.

Extra baggage and a limited amount of freight will
be taken at reasonable rates.

It will be required that each person upon entering
his name for the expedition, pay down $50 – the bal-
ance to be paid previous to the starting of the train.

NO APPLICATIONS WILL BE RECEIVED AFTER MARCH
15, 1859.

As the number will be limited, persons desiring to
connect themselves with the expedition must make
immediate application (by letter or otherwise,) to S. O.
HEMENWAY, proprietor Rowena House, Lecompton,
Kansas. [References are listed]

FREIGHTING TO PIKE'S PEAK [329]

Transportation of baggage, provisions and merchandise.

The undersigned, having made arrangements for
transporting freight for emigrants to Pike's Peak, will

[329] Advertisement in the *Missouri Republican*, February 26, 1859. The
Nebraska City News of February 26 gives reports from many towns telling of
parties leaving, or about to leave for the mines.

send out during the season, fifty trains of twenty-six wagons each – or more if necessary – from Westport, Missouri, and from Atchison, Kansas territory. This will afford an opportunity for merchants and emigrants of having their provisions, merchandise and other freight, transported for a stated price per 100 pounds; and at prices much less than private teams can be had. I will also carry passengers at a stated price, furnishing board and transporting their baggage. Messrs. Samuel & Allen are fully authorized, from this date, to make contracts for transportation of men, merchandize, etc. Freight will be received by them at St. Louis, and by myself, at Westport, Missouri, or at Longwood, Pettis county, Missouri, and at Atchison, Kansas territory, or from Messrs. Samuel & Allen, No. 132 Second street, St. Louis. JOHN S. JONES [330][References listed]

GOLD WASHERS [331]

Our fellow citizen, Mr. Masuch, (gunsmith on Delaware street,) has got up a new gold washer that we must confess is the most complete, useful and unique machine for washing gold that we have ever seen worked. We are not mechanic enough to attempt to give a minute description of its mechanism; we can only say that it is far ahead of any washer ever used in California. As a test of its worth, we will say that Mr. Masuch took a $5.00 gold piece and filed it up into as small particles as he possibly could. He then took and mixed the filings with five bushels of earth, and run the earth and gold through his washer. On cleaning out his machine, sifting and blowing his dust,

---

[330] Jones and W. H. Russell were presently to launch the Leavenworth and Pike's Peak Express.

[331] Kansas City *Journal of Commerce*, February 27, 1859. The *Hannibal Messenger* of the same date says the tide of emigration for the mines is beginning to flow through Hannibal.

he found on weighing his gold, that not one particle had been lost.

A machine that will do this, is just what the gold hunter wants, and we must advise everyone on their way to the Peak, to call and see Masuch's gold washer.

OUTFITS FOR THE NEW GOLD DIGGINGS [332]

Colt's revolving pistols and rifles at reduced prices. Also, "Sharp's", "Allen's", "Green's" and "Hall's" breech loading rifles and carbines. MINNIE RIFLES, all kinds; DOUBLE RIFLES; DOUBLE RIFLE AND SHOT combined.

SINGLE AND DOUBLE SHOT GUNS of every kind and description, from common iron barrel to finest laminated steel.

PISTOLS of every variety, single and double barrel. Allen's revolvers, Allen's breech-loading pocket and rifled-belt pistols; money belts; pocket compass; Ames' shovels, pickaxes, etc., etc., etc. WILLIAM READ & SON, 13 Faneuil Hall Square, Boston

EMIGRATION AND A NEWSPAPER PRESS [333]

Emigration to the gold region of western Nebraska and Kansas, is already beginning to take place. Several small parties have started from this city, within the past two weeks, and numbers of others are preparing to go shortly. . . A large train of hand carts is expected to go out from this city. It has some advantages over the more common mode of going with ox teams, and is generally adopted by the Mormons, numbers of whom leave this city and other places in the vicinity every spring for Salt Lake. The expenses of the trip

[332] Advertisement in *Boston Daily Journal,* February 28, 1859.

[333] Correspondence of February 28, from Omaha, and published in the *New York Tribune* of March 19, 1859.

need not exceed $25, and the emigrant is free from many of the difficulties attendent upon teams, such as losing cattle, sticking in sloughs, crossing streams, etc. . . .

Several saw mills are already on the way; and a company of persons from this city will start this week with a press and materials for printing a newspaper, to be called *The Rocky Mountain News,*[334] at Fort Laramie or some other suitable point in or near the mining region. The first number is expected to be issued about the first of April.

The accounts which continue to come in from the miners are not so uniformly favorable as at first. . .

[BOATLOADS FOR THE MINES] [335]

. . . Well, the long looked for immigration has commenced. Friday, Saturday, Sunday and Monday, have each brought its boat, and each boat its load of gold seekers. . .

A party left here yesterday on foot for the mines with a hand cart ingeniously contrived to ford all the streams like a boat. Another large train leaves tomorrow. . .

Capt. Samuel Cooper, our very efficient chief of police, handed in his resignation today. He starts in a few days for Pike's Peak.

[CHARACTER OF THE '59ERS] [336]

If the shiftless, lazy, lousy, scurvy, profane, insane and idiotic herd of rapscallions, nincompoops and

---

[334] The first issue of this pioneer newspaper was published at the mouth of Cherry creek by W. N. Byers on April 23, 1859. It is interesting to learn that the name of the paper was chosen so early.

[335] Leavenworth correspondence of February 28, appearing in the *Lawrence Republican* of March 3, 1859.

[336] An editorial in the *Nebraska City News* of March 3, 1860. In its issue of April 14, 1860, the editor comments further: "The scurvy horde of pauper

ninnies, make their way up the Platte in the months of March and April that found themselves rushing headlong, they knew not where nor whither, in these same months last year [1859], like the mad king of France who "marched up the hill then marched down again," we think the gold product, for them at least, will be very small.

PIKE'S PEAK [337]

Persons going to the gold regions can have their lives insured, either for the benefit of their families or creditors, by applying to

T. E. COURTNAY
Agent for the
New York Life Insurance Company,

AT HIS OFFICE
N. W. COR. MAIN AND OLIVE STS.
[Irrelevant parts omitted]

emigrants who started for the mines last season, but whose courage for the most part oozed out about fifty miles west of this city, will not soon be forgotten in this city. The same amount of reckless folly, ignorant presumption, and crazy madness were never before witnessed in our city or on the great plains of the west.

An equal amount of raw material, of pauperism, and ignorant idiotcy have never before congregated here; and if the poor houses and brothels of the north and east should vomit forth their contents, we question whether they would not sicken at the sight of the imbicility and vagabondism exhibited on this frontier one year ago. . .

Thank Heaven no such tom foolry is witnessed this season. To all appearance a far different class is this year making their way to the land of gold.

Solid men of business, possessed of sagacious skill and prudent judgement, with machinery and capital, earnest hearts and force of character, are this season making their way to the mines, determined to develop, in part at least, the inexhaustible wealth of that rich mineral country. We are glad to see so good and so respectable a class of citizens emigrating to the rich gold fields in the western part of our territory.

It bespeaks and betokens great things for the country.

We congratulate the new territory of Jefferson upon the brilliant future that is just dawning upon her."

[337] *Missouri Republican,* March 5, 1859. The *Leavenworth Times* of March 5 tells of four young men, Messrs. Cook, Keller, Reese and Mackay, having left for the mines with a handcart.

OFF FOR THE DIGGINGS [338]

Not a day passes that we do not witness the departure of companies for the new El Dorado to the west of us. Over seventy-five companies, averaging six men each, have outfitted here within the last two weeks, and are now "on their winding way" to Cherry creek. The number of companies that have outfitted at Council Bluffs and other places and crossed the river at this place, may safely be estimated at seventy-five more.

The equipage of these companies are as varied and different as the character and means of the individuals composing them. Numbers have left with heavy wagons drawn by four horses or mules, others with three or four yoke of oxen, some with a pair of mules, not a few with a single horse, and a great many with hand carts, while others have started on foot and alone. . .

FOR PIKE'S PEAK [339]

The current through our city to Pike's Peak, is widening and deepening daily. The stage company acknowledge it; having companies almost nightly, to enjoy their hospitality. Our merchants make mention of the fact, with a merry twinkle of the eye, and an exuberant feeling extended clear to the depths of their pockets. Of course we are only witnessing, as yet, the few drops that precede the copious shower. . .

THE NEW EL DORADO – EMIGRATION REALLY COMMENCED [340]

Those two little words, "Pike's Peak," are every-

[338] *Omaha Nebraskian,* March 9; reprinted in the *Wyoming Telescope,* March 19, 1859.

[339] From the *Des Moines Citizen* and reprinted in the *Council Bluffs Bugle* of March 9, 1859.

[340] *Missouri Republican,* March 10, 1859. The *Missouri Democrat* of March 11 carries a story of two men setting out from Springfield, Missouri, with a wheelbarrow, one man pushing and the other pulling.

where. The latest from Pike's Peak is eagerly devoured, no matter what it is. The quickest, safest route to Pike's Peak is what thousands want to know. Pike's Peak is in everybody's mouth and thoughts, and Pike's Peak figures in a million dreams. Every clothing store is a depot for outfits for Pike's Peak. There are Pike's Peak hats, and Pike's Peak guns, Pike's Peak boots, Pike's Peak shovels, and Pike's Peak goodness-knows-what-all, designed expressly for the use of emigrants and miners, and earnestly recommended to those contemplating a journey to the gold regions of Pike's Peak. We presume there are, or will be, Pike's Peak pills, manufactured with exclusive reference to the diseases of Cherry valley, and sold in conjunction with Pike's Peak guide books; or Pike's Peak schnapps to give tone to the stomachs of overtasked gold diggers; or Pike's Peak goggles to keep the gold dust out of the eyes of the fortune hunters; or Pike's Peak steelyards (drawing fifty pounds) with which to weigh the massive chunks of gold quarried out of Mother Earth's prolific bowels. . .

The Pike's Peak fever, added to the usual influx of country merchants at this time of the year, have served, as we remarked yesterday, to fill the city with strangers. The trains of the rail-roads terminating here bring large numbers every trip; all the hotels are crowded, and St. Louis at present wears the appearance of a popular overflow. At such a time rogues find their harvest. . .

At this moment there are hundreds of professional pick-pockets, pigeon droppers and "confidence men" here, ready to take all the cash and valuables brought by the strangers visiting the place. The levee is thronged, and the Missouri river boats particularly are beset by them from morning till night. The police force is wholly inefficient, or perhaps we ought to say insufficient, to be of much service. . .

WHITE TOPPED WAGONS! [341]

Our streets during the week have been lined with white topped wagons. Many of them conveying families and settlers to the broad and fertile prairies west of us. The vast majority are gold seekers for western Nebraska and Kansas.

EN ROUTE TO PIKE'S PEAK [342]

St. Joseph, Mo. March 12, '59

Editors Press and Tribune [Chicago, Ill.]: I arrived here on the 10th, after a tedious trip of three days. There are hundreds coming on every train and hundreds are here now.

St. Joseph is a perfect jam with "Peakers" and sharpers takin 'em in, horses, mules, oxen, men, women, children, wagons, wheelbarrows, hand carts, auctioneers, rumors, stool-pigeons, greenhorns and everything else that you can imagine, and a thousand other things your imagination will fail to conceive. Everything is very high; board at a "one horse" hotel $2.50 per day, and little rats of mules $150. The folks think the whole United States will be here in a few days. Ten days ago a man could fit out here at a reasonable rate. There are hundreds starting from here, but they are the poorest of creation. I would not have believed it, but it is a fact that there are hundreds now starting on foot, with nothing but a cotton sack and a few pounds of crackers and meat, and many with hand-carts and wheelbarrows. . .

---

341 *Nebraska City News,* March 12, 1859.
342 *Wyoming Telescope,* April 9, 1859.

"PIKE'S PEAK" [343]

We notice any quantity of wagons passing through our streets bearing this significant inscription upon their side boards.

EMIGRANTS FOR PIKE'S PEAK [344]

During the present week the river has been crowded with boats, and the boats with passengers destined for Pike's Peak. The number stopping at this place to get their outfit will number about seventy-five persons to the boat – perhaps more – but we desire to fall within the number.

The hotels in the city are crowded. . . Gold rockers, black carpet bags, picks, pans, oxen, mules, and provisions are all the rage. . .

PIKE'S PEAK [345]

It is astonishing how rapidly we learn geography. A short time since, we hardly knew, and didn't care, whether the earthly elevation called Pike's Peak was in Kansas or Kamtschatka. Indeed, ninety-nine out of every one hundred persons in the country did not know

[343] *Nebraska City News,* March 12, 1859. In the same issue of this paper we read of several companies from Michigan outfitting in Nebraska City. Samuel S. Curtis and D. F. Richards arrived at Omaha on March 11, and the next day the *Omaha Republican* issued a special edition giving their report. They said they had left the mines on February 16, at which time little mining was being done. But a "small number of persons were at work with rockers, working from two to five hours a day, as the weather would admit, and making from two to five dollars per day."

[344] *Kansas Weekly Herald* (Leavenworth), March 12, 1859. The same paper carries the detailed terms (sixteen numbered paragraphs) of the great freighting firm of Russell, Majors, and Waddell, for carrying freight to Denver. Jones and Russell propose to carry baggage for gold seekers, from Leavenworth to Denver, at the rate of twenty-five cents per pound. The men are to walk.

[345] *St. Louis Evening News,* March 17, 1859.

that there was such a topographical feature as Pike's
Peak. Now they hear of nothing, dream of nothing
but Pike's Peak. It is magnet to the mountains, toward
which every body and everything is tending. It seems
that every man, woman and child, who is going anywhere at all, is moving Pike's Peakward.

On the streets of the city, you see nothing but "Pike's
Peak" flauntingly blazoned on every fabric of iron,
wood, wool or cotton, that a mortal is presumed, in
any of the exigencies of life, to stand in need of. Every
one who has anything to sell puts it in the current of
Pike's Peak trade, and it is carried off in a jiffy. Along
the sidewalks, at the outfitting stores you will see complete sets of out-of-doors furniture – tripods with the
camp-kettle swinging over the place where the fire
ought to be, and ready in a moment to furnish the
Pike's Peaker with a cup of coffee; supple-jointed con-
trivances that look like little bundles of sticks, but
which, when opened, expand into a virtuous bedstead
large enough to contain a slumbering six-feeter, on
his way to the "Peak;" frightful looking bowie knives
manufactured for the special purpose of disembowling
the Indians around Pike's Peak; many-muzzled pistols
and guns, intended to be used by one Pike's Peaker,
when he proposes to let a little daylight through the
proportions of another; hats made with the sole view
of protecting the head of the miner from the Pike's
Peak weather; big boots for the Pike's Peaker to "slosh
about" in the mud and water with; medicine for the
Pike's Peaker to use when he gets sick – and, in short,
all the articles that a considerate and careful world
can conjure up to minister to the Pike's Peaker's
comfort.

EMIGRATION TO PIKE'S PEAK [346]

We noticed a train of very unique character wending its way Saturday, through our city, for the gold mines. We did not count them, but they had hand carts, very light and neat and painted red. These, containing clothing and provisions, were drawn, each, by two men, while several others followed two by two. We learned that eight men were alloted to each cart.

EN ROUTE TO THE GOLD MINES [347]

Leavenworth City, March 17, 1859.

(Correspondence of the *Republican*.) Well sir, here we are, this far on our road to the wonderful fields of Kansas. Since we left St. Louis nothing very astonishing has occurred, unless it was the remarkable facility with which a couple of flashily-dressed gentlemen cleaned out our party at a small game of poker on board the boat. With this exception, our journey was a pleasant one. We were neither sunk nor blown up, and yesterday landed at this place safe and sound. Beyond all question this is the most growing and thriving town west of St. Louis. At present, the great throng of Pike's Peakers, as we are facetiously called, which crowd her streets, gives the place a very business-like appearance, but independent of this, she bears internal evidence of life and vitality. . .

Those who are flush purchase cattle, mules and wagons, and go well provided with all that is necessary to make a trip on the prairies with comfort and pleasure. The next class takes the hand cart and wheelborrow, while the poorest, and, I fear, the most numerous,

---

346 *Nebraska Advertiser*, March 17, 1859; reprinted from the *St. Joseph Gazette*. The Kansas City *Journal of Commerce* of March 15 says that 140 passengers (mostly for the gold region) had landed at Kansas City during the preceding forty-eight hours.

347 *Missouri Republican*, March 23, 1859.

take it on foot. The hand-cart appears to be quite a fashionable vehicle with the "Peakers," as it is cheap and has a decided advantage over a knapsack or carpet-bag. The Mormons have reduced this mode of travel to a science; they have practiced it for years, and large trains of carts pass every year from the Missouri river to Salt Lake City, a distance of 1,200 miles, with apparent ease and celerity. But then each train of forty or fifty carts is accompanied by four or five ox wagons, carrying tents, provisions, etc. . . Yesterday I saw some men starting out with their mining shovels over their shoulders and their diminutive carpet bags on the end of them. There were not five day's provisions in the whole party. One party of sixteen or twenty started with one old horse, and fifty pounds of hard bread. Their intention was to kill game on the road and to sleep in barns at night. They appeared to think that the prairies were covered with barns and sheds, built by the Indians to shelter the buffalo. . .

Since my arrival here, however, my faith has been considerably shaken in the Kansas gold mines. I have met some people who say that the whole thing is a grand humbug, gotten up by persons interested in embryo towns on Mr. Gilpin's "plateau," and by merchants along the Missouri river who have goods to sell. Be this as it may, the Pike's Peak crowd has got up steam for the trip, and nothing can stop it. . . A hopeful and sanguine PIKE'S PEAKER.

[RUSSELL'S PARTY FROM GEORGIA] [348]

The *Dahlonaga Signal* says: Mr. William [Green]

---

[348] *Columbus Daily Sun*, March 18, 1859. Russell was the leader of the party that made the gold discoveries in the summer of 1858. The *Macon Journal and Messenger* of March 30 quotes the *Memphis Avalanche* as saying that this party, consisting of 49 men, mostly from Georgia, took the steamer at Memphis on Monday for St. Louis.

Russell, of Dawson county, requests us to say that he, in company with 75 or 100 persons, will start for the gold mines of the Rocky mountains on the 15th inst. On the 20th he will be at Dalton without fail, at which point he will be pleased to meet and welcome all who desire to try their luck in the newly discovered mines. The distance is near two thousand miles, and provisions, implements, etc., will be laid in at Leavenworth, K.T. An outfit will range from $125 to $150. We wish our friends every success. Mr. Russell adopts this course to subserve the answering of numerous letters received.

RIVER NEWS [349]

. . .Business on the landing was brisk, considering the state of the weather, and the cheerless prospect for better. Pike's Peakers and other emigrants continue to besiege the landing, and Missouri river boats. Three boats arrived from the Ohio river, crowded, and one in the same condition from Illinois river. They are coming from all quarters of the country. North, south and east, are pouring their foot loose population directly through our city, en route for the gold regions. . .

FOR THE MINES [350]

Upwards of seven companies left our city Tuesday for the mines. Some were well provided and will have a pleasant trip, others had a moderate outfit and will probably get through in safety.

But one company embraced an amount of foolhardiness we are pained to record. This company consisted of sixteen able bodied fellows with blankets, picks and pans strapped to their backs. Their entire lot of provisions consisted of forty pounds of crackers and a

[349] *Missouri Republican,* March 19, 1859.
[350] *Leavenworth Times,* March 19, 1859.

quantity of salt – the latter being barely sufficient to preserve the former in case it was not eaten.

On being asked how they expected to make the trip of 500 miles with their ridiculous outfit, one of them replied, "That's easy enough. We intend to kill enough game and sleep in barns."

Verily the fools are not all dead.

NEW AND GREAT FREIGHT LINE [351]

We have received the circulars and advertisements of the "Kansas City Gold Hunter's Express Transportation Company," which has been fully organized. The style of the firm is Irwin, Porter & Co., and embraces the oldest and most experienced freighters on the plains. Their arrangements both here, at St. Louis and in the eastern cities, are such as to make it the most reliable and safest company on the plains, and their business will be more thoroughly systematized, and under charge of more experienced men, throughout, than any similar organization ever known in the west. . .

EMIGRANTS [352]

For the last week we have had one or two boats arriving every day, with crowds of emigrants for Pike's Peak, most of them stopping at this place [Leavenworth]. Hundreds have already arrived here and started out on the plains, and there are hundreds still in the city.

SCRAPS BY THE TRAVELER – HO! FOR THE MINES [353]

Kansas City, March 21, 1859.

Here they come by every steamboat, hundreds after

351 Kansas City *Journal of Commerce,* March 19, 1859.
352 *Kansas Weekly Herald,* March 19, 1859.
353 *Missouri Republican,* March 27, 1859.

hundreds from every place – Hoosiers, Suckers, Corn crackers, Buckeyes, Red-horses, Arabs and Egyptians – some with ox wagons, some with mules, but the greatest number on foot, with their knap-sacks and old-fashioned rifles and shot-guns; some with their long-tailed blues, others in jeans and bob-tailed jockeys; in their roundabouts, slouched hats, caps and sacks. There are a few hand-carts in the crowd. They form themselves into companies of ten, twenty, and as high as forty-five men have marched out, two-and-two, with a captain and clerk, eight men to a hand-cart, divided into four reliefs, two at a time pulling the cart, which contains all the provisions, camp equipments, working tools, et cetera, for the eight persons thus arranged. Onward they move, in solemn order, day after day, old and young, tall and slender, short and fat, handsome and ugly, the strong and the weak, the red-faced rumbruiser and the lean, lank Jonathan, with grimmage sour, striding among saints and sinners, heads up and still upper lips, march forward as "lords of creation," and seeming "monarchs of all they survey." Nothing daunted, and nothing short of the trip will satisfy their mighty imaginations, air castles and visionary hopes of immense wealth they are to receive for their pains. Enthusiastic, merry, with light hearts and a thin pair of breeches, they calculate to accomplish all their fondest hopes. Many have sold out all their homes, all their valuables, to furnish themselves with an outfit for Pike's Peak mines. Others left their wives and children behind with no protecting arm or support but what Providence may give, and blindly rush headlong into the wild delusion of glittering sands full of golden eggs. 425 arrived on the *Alonzo Child* to-day.

The runners and drummers for hotels and express wagons are thick as hail, with the cry, "$40 for passage

to Pike's Peak from here; charge $50 at Leavenworth."
"Shortest route, best roads, fine eating houses, good
camping grounds – only 550 miles." Some stop, while
others go on to Leavenworth, and take the Junction
City road. The city of Leavenworth is full of them,
who tarry but a little while after landing, fall into line,
and take their march onward, without expending a
solitary dime. Most of them are entirely out of money,
but their means are invested in the mess – their money
is all to come from the mines. Poor deluded creatures,
running after shadows and phantoms, and bitter will
be their cup when riches have taken wings – out of
provisions, away from home, sick, sore and disappointed, they look homeward to cheerful faces and
happy little ones, and exclaim, "no balm in Gilead."
Only think of 100,000 persons scattered over a few
miles of territory, digging and tearing up the ground
in search of wealth.

Eating greasy meat, sleeping on the cold ground,
in rain, snow and hail, hot and cold, wet and dry – up
early and late – striving like giants to turn upside down
the earth for the sake of gain, away from the cheering
hearthstone, joyous laugh and happy home, among
wolves and savage beasts, working out their lives for
naught, when, with one half that energy, industry and
perseverence, at home on their farms, they would have
full pockets, good health, be with their family and
friends, enjoying happiness, comfort, and peace. EZEL.

MORE FOR THE MINES [354]

The *Alonzo Child* brought up a large delegation for
the gold mines. We saw all sorts of conveyances – carpet sacks, wagons, carts, hand-carts, ox teams, mule
teams, human teams – all busy at work preparing to

[354] Kansas City *Journal of Commerce,* March 22, 1859.

start. One conveyance struck us as peculiar. It was a cart with a body decked over, and bows fixed like to a wagon bed. The inside is abundantly roomy for provisions, and the deck is intended for a sleeping bunk. Thus equipped, it answers all the purposes of a portable house, and with its tandem team, is bound to go through in good style and in good time.

AMBULANCES FOR THE MINES [355]

Messrs. Wiley & Harris, coach makers, McGee's Addition, are constructing ambulances for the mines, that are just the thing. In addition to being light, strong and roomy, the seats are made to unfold so as to form a complete bedstead, on which a mattress may be fixed up for a king – all under a good oil covering, obviating all necessity for a tent.

THE LEAVENWORTH AND PIKE'S PEAK EXPRESS COMPANY [356]

(Leavenworth City Correspondence.)

Leavenworth City, March 23, 1859

. . . A number of leading representatives of the business community of this city, concluded in the early part of February last, to associate themselves for the purpose of creating a company for the transportation of passengers and freight to the mining districts with the greatest possible safety and dispatch. In due course of time the organization of a company for the purposes already named was effected and completed by the subscription and cash payment of stock to the amount of two hundred thousand dollars, and the election of

[355] *Ibid.*, March 22, 1859. Another story in the same paper says: "The gold hunters are crowding the city by every steamer. The arrivals since Saturday night have been over two hundred, and every boat is black with people."

[356] Published in the *Missouri Republican*, March 28, 1859. This company established the first stagecoach line to Denver, the first coach arriving May 7, 1859. See L. R. Hafen, *The Overland Mail*, chapter VII.

Mr. Wm. H. Russell, the famous government freighter as president, and of Mr. John S. Jones, of Pettis county, Missouri, the pioneer government contractor of the west, as superintendent. To the latter gentleman the company, knowing that his ability and experience would be more than equal to the management of even so large an enterprise as that in contemplation, very wisely delegated discretionary powers. The capital of the company is represented by 40 shares of $5,000 each, the whole of which is now held by ten individuals. It can be increased as the wants of the company demand it. The company adopted the name of "Leavenworth City and Pike's Peak Exportation [Express] Company." The familiarity of Mr. Jones with the topography and the resources of the plains, enabled him to make the best possible selection of a route that was fully adapted to the objects of the company. . .

For the present, Sibley tents will be provided for the accommodation of passengers and store-keepers at each station. As soon as the necessary material can, however, be brought to the various points, permanent and solid structures will be erected. . .

The first train for the supply of the several stations with men, stock, provisions, etc., will leave on the 25th, and will be succeeded daily by other trains until the entire length of the line will be fully equipped, which is to be previous to the 10th proximo, on which day the first passenger coach will leave Leavenworth City.

The fast passenger line will be supplied with not less than fifty of the celebrated Concord coaches, and eight hundred mules.

The managers of the company will warrant, from the very start, a transit in less than twelve days. The line once being in full working order, they will run the entire distance in less than eight days. . .

RIVER NEWS [357]

Still they come! Who come? The Pike's Peakers and gold seekers – fortune hunters. Every boat from the Ohio is crowded with them, and every boat for the Missouri goes up jammed with people. Boats now loading for this port and the Missouri river, at Pittsburgh and Cincinnati are thronged with emigrants. Those on their way here are in the same condition. They are coming down the Mississippi and Illinois, too – the Pike's Peakers. They are flocking overland, across the country – the Pike's Peakers – all with strong limbs and brave hearts, capable of standing the exposures of the prairie wilderness and mountains. . .

PIKE'S PEAKERS [358]

If you see a man come up the river, or get off a steamboat, or walking along the streets with an india rubber coat on, set him down for a Peaker. The city is full of them at this writing and every boat brings more. We saw one hotel runner heading a company of twenty-five yesterday morning up Main street to his hotel, with all the "pomp and circumstance of a glorious military captain." The Santa Fe road is lined with them, and on all hands the cry is, Pike's Peak-gold-outfit, etc. This is a great country.

[KANSAS] [359]

. . . As for the country, the land is as cheap as dirt, and good enough; but the climate is rainy, blowy and sultry. The people die so fast, that every man has his third wife, and every woman is a widow! They fulfill

---

357 *Missouri Republican*, March 25, 1859.

358 Kansas City *Journal of Commerce*, March 26, 1859.

359 Extract from a Kentuckian's letter to his wife, published in the *Hannibal Messenger*, March 27, 1859.

the Scriptures to a letter, where it says, "Let God be
true, but every man a liar!"

RUSH FOR THE MINES [360]

Hundreds of persons pass through our city daily en
route for the new gold mines. We believe the emigration to Pike's Peak this season will be far greater than
the number that went to California in any one year.
If the emigration continues at the present rate during
the summer, we believe there will be at least seventy-
five thousand persons at Pike's Peak by fall. . .

We are daily in receipt of news from the mines,
while some is very favorable, we occasionally receive
news not so flattering. We shall endeavor to give both
sides, that our readers may judge for themselves.

PIKE'S PEAK EMIGRATION [361]

. . . If 100,000 men should go to Pike's Peak this
season (which is probable) from all sections of the
union, it will take the snug little sum of $2,000,000
to furnish them a fitting out at $200 per head, which is
a light estimate.

CONFLICTING REPORTS [362]

During the past week we have had several arrivals
and letters from the gold region, and as many conflicting reports as to the riches of the mines.

On Monday last two Germans who formed part of
the German company that left this city for the mines
last fall, arrived in Atchison. One of them says that
the mines are a humbug, and that no one can make

[360] *Hannibal Messenger,* March 29, 1859.
[361] *Lawrence Republican,* March 31, 1859.
[362] *Lawrence Republican,* March 31, 1859.

enough to pay him for his labor. . . The other German says that the mines are rich, and that gold can be found in any place, and in quantities sufficient to pay a man with merely a pick and pan from $2.00 to $5.00 per day steadily, with the prospect of striking rich "leads," from which $15.00 to $20.00 per day have been taken. He says that very few are dissatisfied except those who are too indolent to work. . .

A PRINTING PRESS FOR THE MINES [363]

Yesterday we fitted out for Mr. T. A. Fullerton, of this city, a press and the necessary material for publishing a newspaper in the mines, to be called the *Cherry Creek Pioneer,* which left in his express during the day. Mr. John Merrick, who was a short time since connected with the *Elwood Press,* is connected with Mr. F. in the publication of the *Pioneer.*[364]

Mr. Merrick left several weeks since with a portion of the type for their paper, and in a few weeks – now that the press has gone – a newspaper will be issued under the shadow of the Rocky mountains, in the midst of the gold bearing placers.

EL PASO MINER [365]

El Paso Miner, a weekly newspaper to be published at El Paso, near the base of Pike's Peak, K.T. Its publication will be commenced in June next by Mr. S. S. Prouty, at $5 per annum; $3 for six months, in advance.

[363] *Missouri Statesman,* April 1, 1859; reprinted from the *St. Louis Journal.* In its issue of March 15th the *Journal* says that Mr. Fullerton and company have established an express line to the mines, and will run an express twice a week. Apparently this never materialized.

[364] The first and only issue of the *Cherry Creek Pioneer* was printed by Merrick at the mouth of Cherry creek on April 23, 1859—the day of the first issue of the *Rocky Mountain News.*

[365] *Missouri Statesmen,* April 1, 1859. This newspaper was never published.

[LIGHTNING EXPRESS][366]

. . . A little car passed here the other day propelled by a little white bull, whose sides were painted thus, "Pike's Peak Lightning Express," and the six jolly native Americans, whose property it was, seemed to be in great good humor with themselves and the rest of mankind. I saw the same party leave Council Grove the other morning, themselves harnessed into the car, having sold the bull, but still called their team the "Lightning Express."

PIKE'S PEAK CORRESPONDENCE [367]

Westport, Mo., April 4, 1859.

The Pike's Peak emigration continues. The loads of men and bacon that one sees every day, en route for the mines, makes one wonder what there is in the discomforts of life so enticing. I saw half a dozen men with one wagon, this afternoon, stuck fast in a mud hole south of town. They have been all day coming four miles, from Kansas City. . . The emigration does not fall off, but it is more respectable in numbers than quality, which cannot be boasted of.

SMOKY HILL ROUTE – SUFFERING [368]

Mr. J. C. Hill, a trader at Council Grove, was in town yesterday. He informs us that before he left home, a party of gold hunters arrived there, who started for the mines by the Smoky Hill route. They found neither road nor trail, and after wandering for fifteen days over the country, and provisions giving out, they struck southward to the Santa Fe road, which they reached

[366] Correspondence in the Kansas City *Journal of Commerce* of April 3, 1859.

[367] *Missouri Republican*, April 10, 1859.

[368] Kansas City *Journal of Commerce*, April 6, 1859.

near Council Grove. They had suffered terribly from storms and cold. After recruiting and laying in provisions, they left again for their destination, taking the Santa Fe road. How often will it be necessary to tell the public that there is no road up the Smoky Hill. A few hundred deaths from deprivation may be necessary to bring them to their senses, and these will not long be wanting.

PIKE'S PEAK AND POISON [369]

Our readers will perhaps be a little surprised when we make the statement that the large emigration to Pike's Peak, has had the effect of raising the price of corrosive sublimate and calomel, to nearly double what those articles were selling at before the breaking out of the gold fever. Not that those valuable medicines have been very largely used in the treatment of this disease, which is daily "carrying off" its thousands, nor that the emigrants are going out with the murderous determination of procuring gold by a more easy method than digging, from those who have already acquired it. The case is solely attributable to that fact, that corrosive sublimate and calomel, are preparations of mercury or quicksilver, which metal is extensively used by miners for separating the gold from the worthless substances, and the advance in price created by the great demand for quicksilver, has given rise to corresponding advance in all mercurial preparations.

---

[369] *Central City and Brunswicker,* April 6, 1859. The *Topeka Tribune* of April 7 tells of two men leaving Topeka with a handcart for the mines. The *Nebraska Advertiser* of the same day says that a handcart train from Oregon, Missouri, passed through Brownville en route for the mines. The Kansas City *Journal of Commerce* of April 7 tells of a wagon, especially prepared for comfortable sleeping quarters, being taken to the mines. The *Atlanta Weekly Intelligencer* of the same date reports the departure of W. D. Moore (printer) and J. L. Ware for the gold region.

[DOG TEAM FOR PIKE'S PEAK] [370]

The Aledo [Mercer county, Illinois] Record says that Mr. R. S. Osbaris of that place, is about starting for the gold regions with a dog train. He has a light wagon, two Newfoundland dogs, two grey hounds and two pointers for the lead, and expects to distance all competition.

FEMALE PIKE'S PEAKERS [371]

The *Pittsburg Journal* notices the departure of quite a number of young ladies from that city en route for Pike's Peak. They have little idea of the hardships they may have to undergo during such a journey.

LETTER FROM KANSAS [372]

Robinson, Brown co., K.T., April 12, 1859.

To the Editor of The Boston Journal: . . . The roads in this vicinity, leading west from the river, are white with the wagons of Pike's Peak emigrants. They are camped in large parties along the waters of almost every creek, waiting for the grass to start before proceeding further. Those who went out a month ago, and are now beyond the timbered region, must have suffered greatly during the recent snow storms and severe weather. Many started with small stocks of provisions that are already exhausted. . . The number in the territory now en route for the mines, is little less than fifteen thousand, and there is no perceptible decrease in the immigration. Meanwhile the reports from the mines continue very conflicting, and though many

---

[370] *Wyoming Telescope,* April 9, 1859. The same paper quotes the *Nebraska Republican* in telling of an emigrant with two large dogs attached to a light wagon in which were his mining tools and provisions.

[371] *Wyoming Telescope,* April 9, 1859.

[372] *Boston Daily Journal,* April 19, 1859.

are sanguine, the information is by no means satisfactory.

The number of hand-carts and wheelbarrows on the way is almost incredible. In St. Joseph, a few days since, a train of six hand-carts, with the emigrants harnessed in, had just started for the ferry, and attracted considerable attention, when a fellow of most solemn visage shouted out to them at the distance of half a square: "Halloa! hold on there." The goldseekers stopped while he came up, and asked: "Are you going to Pike's Peak?" "Yes," was the rather crusty response. "Well, why don't you wait for the grass?" continued the interrogator. "Grass!" ejaculated one of the emigrants impatiently. "What do we want of grass? we haven' any cattle." "Very true; but you are making asses of yourselves, and you ought to look out for provender!" . . . A. D. R.[Richardson]

INTERESTING FROM THE FRONTIERS [373]
Kansas City, Mo., April 13, 1859.
TO THE EDITOR OF THE BOSTON JOURNAL: . . .
Hundreds of eastern immigrants continue to arrive and depart daily for Pike's Peak. Though a good many glowing accounts are coming in from the gold region for publication, I find a strong undercurrent of adverse representations. Those who are sanguine of success base their opinion on what they hope for the future, rather than what is actually being done. Gold is not arriving at the border towns in sufficient quantities to justify the belief that the mines are now paying. . .
A. D. R.

[373] *Ibid.,* April 20, 1859.

[WIND WAGON] [374]
Westport, April 14, 1859.

What do you think of a prairie ship? Well, there is one in this town, but they call it a "wind wagon." It is a queer looking affair, and I was forcibly struck with the picture it presented, on visiting it yesterday, and thought at once, of Don Quixotte and the windmill. The affair is on wheels which are mammoth concerns, about twenty feet in circumference, and the arrangements for passengers is built somewhat after the style of an omnibus-body. It is to be propelled by the wind, through the means of sails. As to the wheels it looks like an overgrown omnibus; and as to the spars and sails, it looks like a diminutive schooner. It will seat about twenty-four passengers. . .

The inventor proposes to make the trip to Pike's Peak and back in twelve days, and travel an average of over 100 miles per day. . . Mr. THOMPSON [Thomas], whose inventive mind gave life to the idea now to be early tested, has long been the proprietor of a windmill, on the border not far distant. The launch is to take place on Monday, a few miles south, on the open

---

[374] Correspondence published in the *Missouri Republican* of April 20, 1859. A correspondent of the *Jefferson Examiner* wrote from Kansas City on April 19, as follows: "I went out to Westport yesterday, to take a look at a wind-wagon, built by a Mr. Thomas, purposely to run from that point to Santa Fe. It is a large nondescript sort of a machine, with a body like a small sloop with both bow and stem cut off, suspended on four double spoked wheels, about eight feet in height, with hollow hubs about the size and appearance of one of Wagoner's beer barrels, with steering apparatus, etc. The whole is surmounted by a mast, some twenty-five feet high, upon which will be suspended two sails, cut yacht fashion. This thing is designed to be propelled by the wind over the level prairie at the rate of 100 miles per day. And to-day has been set for the first trial trip. Mr. Thomas, the maker of this wind-wagon, is very sanguine of success, and has been at work at it for five or six years past, and at last completed it, at a cost of $800, through the liberality of some of the citizens of Westport."

prairie. It is said that hundreds will be present to witness the extraordinary sight of a wagon propelled by the wind. The thing is certainly novel, whether it be ephemeral or practical. . .

FOR THE GOLD MINES [375]

Notwithstanding but little efforts were made by our business men, property holders and others to secure travel from our point to the mines, we have had our proportionate share. While other points have had their runners, advertisements and hand bills overrunning the east, giving inflated reasons for the selection of their points as ones from which to start to the mines, we have moved along noiselessly, depending entirely on our merits, geographical position, etc.

The past week has been rather more than usually lively on account of more trains leaving; more have left than in any two weeks previously; and there are others fitting out, and still others arriving daily, both by river and overland. . .

PIKE'S PEAK CORRESPONDENCE [376]

Westport, April 15, 1859

The Pike's Peak emigration, notwithstanding unfavorable reports, is about the same. You can [not] have a better estimate of the gold fever from this: a few days ago I observed a party en route for the mines. The man was driving an overloaded wagon, while the woman was hauling the sick child in a hand-cart, carrying the medicine in her hands. In their hurry to reach the land of gold, they could not stop long enough for their invalid offspring to recover its shattered health. H.C.P.

[375] *Nebraska Advertiser,* April 14, 1859.
[376] *Missouri Republican,* April 21, 1859.

DEPARTURE OF THE PIKE'S PEAK EXPRESS [377]

Leavenworth, April 18.

The first overland express consisting of two passenger coaches, left here at nine o'clock this morning for Denver City. The time through will occupy from ten to twelve days until this line is in thorough working order when it will be reduced to eight days.

The company carries the U.S. mail by authority of the post office department.

The departure of the coaches was witnessed by a large concourse of people. This enterprise is considered one of the greatest of the day and will place us in direct and reliable communication with the mineral region and afford in a short time information of the most conclusive character of its extent and richness.

Two coaches will depart daily hereafter.

Dates to March 19th have been received from Denver City, but they contain nothing not already published.

There were some small receipts of dust here yesterday, and considerable shipments of quicksilver were made today to the mines.

PIKE'S PEAK RUSH [378]

The rush for Pike's Peak continues without any decrease in the number of those bound that way. Sunday the "Pike" and "Fannie" both came down from Quincy loaded with gold seekers and their plunder. Yesterday they left here on the cars for St. Joseph, having laid in their entire outfit at Chicago. There were twenty-seven car loads, including emigrants, cattle, and their outfits. The majority of this crowd were from Chicago; some from Michigan, and others from New Hampshire.

---

[377] Telegraphic news in the *Missouri Republican* of April 19, 1859.
[378] *Hannibal Messenger*, April 19, 1859.

From the Mines [379]

Hampton L. Boon and John H. Ming, of the firm of Ming & Cooper, arrived in our city yesterday morning from Cherry creek, bringing with them seven hundred dollars worth of dust, which was dug upon the Platte, Dry creek, Boulder creek, and in the mountains. Mr. Ming will immediately load his wagons with a general stock of groceries and provisions, and Mr. Boon will buy a large number of milch cows, when they will return by the old Santa Fe road.

Pike's Peak Correspondence [380]

Westport, Missouri, April 19, 1859

. . . It is understood that the wind wagon is to "weigh anchor" in a day or two, and "sail for Pike's Peak." I wouldn't like to be one of the first passengers.

The weather has turned warm at last, but no one expects to see it remain so, but a day. Vegetation is extremely backward. The prairies look black as December, and the trees are as leafless as January. P.H.C.

[Letter of D. C. Oakes] [381]

O'fallons Bluffs, April 22, 1858.

Dear Olive [Mrs. D. C. Oakes]: We are making rather slow progress on our journey We have had very

---

379 Kansas City *Journal of Commerce*, April 19, 1859.

380 Correspondence in the *Missouri Republican*, April 25, 1859.

381 The original of this letter is owned by the State Historical Society of Colorado. Oakes wrote from Plum creek (in the pineries south of Denver, where he had set up his sawmill) to his wife on August 4, 1859. In this letter he said: "This expedition of mine is one that I will long remember for it has tried my fortitude to the foundation on several occasions. While I was on the road I was branded as an infamous villian, one of the getters up of the humbug. My life threatened by thousands on one side, and on the other hand some of my friends pointed out the utter folly and madness of coming through, that I was ruining myself and those connected with me, and to tell you the truth for many nights I slept but little, not from fear of death for that gave me no uneasiness, but it was the thought of what would be heaped upon me

bad weather nearly all the way Several snow storms and very cold and windy. Doc and Aron and myself are going to start on in the morning and leave the ox teams to take thare time. the horses stand it very well so far and the oxen are doing verry well now Some of them are rather weak, but the grass is getting better every day. Nothing of interest has occured on the road except there is a good many turning back before they get through which I think they are very foolish for doing when they get this far there has several men come in from the mines bringing rather discouraging news but I think those men have not given it a fair trial. They have got discouraged before spring. I have considerable talking to do. With one set that was going back, I offered to bet my whole outfit mill and all that I could make $10.00 per day at digging gold there that rather stumped them and hushed them up but they went on home. I want to hear from you and the children very much. Write as often as you can. Yours truly, D. C. OAKES
To Olive M. Oakes

[HANDCART COMPANY COMES BACK] [382]
One of the hand cart trains that left this place a week or two ago, came back on Saturday last, having seen the "elephant" in that mode of traveling to their entire satisfaction. To play horse in dragging one of these carts over the prairies for a week with such weather as we have recently had, will take the "wire edge" off most any one. We always thought it the hight of foolishness for men to start to the mines in this

if the expedition proved a failure. But come through I would and come through I did and humbug or no humbug the mill is paying mighty well now and I think will continue to as long as it is run. . ." See the preceding volume for Oakes's contribution to the Luke Tierney guidebook.

382 *Nebraska Advertiser*, April 21, 1859.

way – not only foolishness, but evincing a lack of good "sound horse sense."

A NEW PHASE OF THE PIKE'S PEAK EXODUS [383]

. . . Recently we have understood that while many of these emigrants have the gold fever well developed, there were hundreds perhaps thousands – these carpet-sack and devil-may-care-boys – who have an entirely different object in view. They start with the intention of bringing up at Pike's Peak, or thereabouts, but they don't intend to stay there. It is now said that, being once there, they can readily drop down upon Sonora and Chihuahua, not exactly as filibusters, but with the intention of taking possession – or at least getting a foot-hold – in those Mexican states. Certain it is, that meetings have been held by them at St. Joseph, if no other place, with the view to consultation about the movement. If they do not find the gold diggings to their liking, then the thousands of idle and disappointed persons about Pike's Peak will be rife for anything, and nothing will be easier than to engage in this preda-tory incursion into the states of Chihuahua and Sonora. Even granting that their inclinations, in many cases, might not lead them to engage in such an enterprise, yet starvation is provocative of many wrong doings, and it will be so with them.

That there are many on the road who seriously con-template this expedition against the Mexican states, we are well assured.

TO PIKE'S PEAK EMIGRANTS – EDITORIAL [384]

What are you going to cross the plains for? To dig gold, eh! Well didn't you know that you could dig for

[383] *Missouri Republican*, April 21, 1859.
[384] *Hannibal Messenger*, April 23, 1859.

gold and get it too, right here in Hannibal? . . . Its
true, we don't call them mines, but by digging in cellars you go through the same operation, and at night
the gold you have been working for is paid you, already in the shape of legal currency. Just look at the
advantages our mines offer you. You are bound to make
an honest living, at any hazard, and in so doing, you
run no risk of being killed by Indians, starving on the
plains, dying from fever, getting your throat cut, or
being turned from honest men to black-legs or gamblers. Don't you see just how the advantages are? If you
are lazy, trifling vagabonds, in mercie's sake, go to
Pike's Peak or Texas, your society will be appreciated
there, while if you curse us with your company, you
might find your way into the – Missouri legislature!

LETTER FROM KANSAS [385]
Sumner, K.T., April 23, 1859.
. . . There is no abatement in the grand hegira for
Pike's Peak. Two daily lines of coaches – morning and
evening – have commenced running from Leavenworth; and the number going out in private trains
seems almost beyond computation. Daily, many a
prairie road is white with their wagons; and nightly,
thousands of prairie slopes are glowing with their camp
fires. The country has witnessed nothing like it since the
California excitement of 1849, when thirty thousand
emigrants crossed the plains. . . Many of the western
emigrants yoke in milch cows with their oxen, thus increasing the power of their team, and providing themselves with a rare luxury; for even in the border towns
of Missouri milk commands 40 cents per gallon, while
whisky sells at 22 cents. The Irish debating society that
discussed the question, "which is the more useful in a

[385] Published in the *Boston Daily Journal*, May 7, 1859.

family, a cow or a barrel of whiskey?" might here gain some light on the subject. . . A. D. R.[Richardson]

PIKE'S PEAK CORRESPONDENCE [386]

Westport, April 25, 1859

. . . A friend sojourning here prior to his departure to Pike's Peak, via Fort Laramie, visited the prairie ship, or wind wagon, to-day. He says that there is not gold enough at Pike's Peak to risk his body in that "awful" looking thing. The inventor is named Thomas, instead of Thompson, and he takes advantage of these windy days to sail his ship a short distance over the prairie. Did the wind always blow one way, and that way favorably; and had Mr. Thomas a track for his "car," similar to a railroad track, then it might run successfully. But so long as the wind blows where it "listeth," and is not particular towards what point of the compass that is; and as there is no track over the plains, but that of the buffalo and the antelope, it is clear to my mind that Mr. Thomas will have to reach Pike's Peak by some other conveyance than the one called his Eolian Car. His ship will ground in some swamp, or break upon some dryland reef. . .

OUTFITTING FOR THE MINES [387]

Several companies of Topeka boys are getting their outfits for a trip to Pike's Peak. Two of the companies leave this morning. Two others will leave on Monday next. The gold fever is growing more violent in this vicinity. Recent reports from the mining region are very favorable.

[386] *Missouri Republican*, May 1, 1859. The Kansas City *Journal of Commerce* of April 21 and 22 carried short articles about the testing of the wind wagon.

[387] *Topeka Tribune*, April 28, 1859.

DISCOURAGING INTELLIGENCE – MINERS RETURNING HOME [388]

In camp, Nebraska City, N.T., April 28th, 1859. Dear Dick: In the wake of the favorable intelligence which I have heretofore reported now comes news of a most discouraging character. . .

These statements I give you for what they are worth, it being impossible, from the contradictory character of the advices, for me to determine anything regarding the matter. . . Truly yours, TALMADGE.

THE GOLD SEEKER'S SONG [389]

Take up the oxen boys, and harness up the mules;
Pack away provisions and bring along the tools;
The pick and the shovel, and a pan that won't leak;
And we'll start for the gold mines. Hurrah for the Peak!

Then farewell to sweethearts, and farewell to wives,
And farewell to children, the joy of our lives;
We're bound for the far west, the yellow dust to seek,
And as we march along we'll shout, Hurrah for Pike's Peak!

Then crack your whips, my jolly boys, we'll leave our homes behind,
And many lovely scenes that we'll often call to mind,
But we'll keep a merry heart, and we'll steer for Cherry creek;
For we're bound to hunt the yellow dust—Hurrah for Pike's Peak.

We'll cross the bold Missouri, and we'll steer for the west,
And we'll take the road we think is the shortest and the best;
We'll travel, o'er the plains, where the wind is blowing bleak,
And the sandy wastes shall echo with—Hurrah for Pike's Peak.

We'll sit around the campfire when all our work is done,
And sing our songs, and crack our jokes, and have our share of fun;
And when we're tired of jokes and songs, our blankets we will seek,
To dream of friends, and home, and gold. Hurrah for Pike's Peak.

[388] Correspondence to the *Davenport Democrat,* and reprinted in the *Missouri Republican,* May 12, 1859.

[389] *Hannibal Messenger,* April 28, 1859.

Then ho! for the mountains, where the yellow dust is found,
Where the grizzly bear, and buffalo, and antelope abound;
We'll gather up the dust along the golden creek,
And make our "pile," and start for home. Hurrah for Pike's Peak.

DISAPPOINTED EMIGRANTS [390]

A party of disappointed emigrants, who left home on the 20th of March last, returned on yesterday. After proceeding out as far as the Big Blue, their faith in the gold mines "gin out," and they took the back track. Their appearance at Market square, with their hand cart and cooking utensils attracted general attention, and created quite a panic among some timid Pike's Peakers, who are now about starting out, while the more knowing ones wanted to know if they "didn't have wives to hum." One of the returned party informed us that he would leave for the east on the evening train, having had enough of the far west, and that as he had just walked six hundred miles, he intended to ride the balance of the way. The west was no place for him.

MUSIC FOR THE MINES [391]

There is now at the Farmer's hotel, a full rigged band of musicians from Indianapolis, en route for the gold mines. They enlivened our streets on Thursday, with the merry strains of music, and are now engaged in preparing their outfits. What a commotion the saxhorns, trombones, kent-bugle and bass drum, will raise among the Kiowas and Cheyennes in the valleys of the Rocky mountains. We admire their philosophy, for music is one of the master keys of happiness.

[390] *St. Joseph Journal,* April 29, 1859, reprinted in the *Missouri Republican,* May 2, 1859. Another report of April 29, published in the *Boston Evening Transcript* of May 16, tells of severe snow storms and great suffering among the gold seekers in the vicinity of Fort Kearny.

[391] Kansas City *Journal of Commerce,* April 30, 1859.

EMIGRATION TO MINES [392]

We received a call yesterday from Mr. N. D. Oaks, of Council Grove, who came in on Thursday. He says the road is literally lined with wagons. He was three days coming in, and as a matter of curiosity counted the wagons for the mines, which amounted to 476. Mr. Oaks says that the men will average six to a wagon, taking hand carts and all other modes of conveyance.

A company of one hundred men came down from the Smoky Hill "route," where they had been lost and exhausted their provisions. They robbed the trading post at the crossing of the Cottonwood, beat the keeper mercilessly, took between 80 and 100 sacks of corn, and all the flour, provisions and groceries on hand, and started for the mines.

Mr. Oaks says that being at Fort Riley a few days since, the ferryman informed him that some days he had crossed from fifty to a hundred emigrants, who had abandoned the "central route," as it is called, and were making due south for the Santa Fe. He also met several parties who had crossed the Kansas at Topeka and Lawrence.

THE PIKE'S PEAKERS. A PICTURE OF THE CROWDS [393]

They continue to come – all sorts, sizes and descriptions. The world seems all a moving. They are passing our office every hour of the day.

There is a fellow from "Illinoy," who wants to know

---

[392] *Ibid.*, April 30, 1859. The *Leavenworth Times* of April 30, tells of the return to Leavenworth of Messrs. Eubanks and Downing, who had been to Denver and back with the reconnoissance party sent out by Jones and Russell to locate a route for the Leavenworth and Pike's Peak Express. They made the round trip in forty-four days, and left Denver April 9. They report good mining prospects.

[393] H. M. McCarty in the *Westport Border Star*, and reprinted in the *Missouri Republican* of May 5, 1859. The *Topeka Tribune* of May 5, 1859 tells of the departure of "twenty-five brave, good fellows, with nine wagons and twenty-seven yoke of oxen."

which is the shortest cut to the Peak, and whether he can get there by Sunday next.

Then comes one all the way from the Green mountains, with a stump-tail mule and a keow with crumpled horns, who, (the mountaineer, not the keow,) had been told that $50 per day could be picked up on Cherry creek – but he guessed it was a darned no-such-a thing. Dew tell!

Then follows a Tennessee rip-snorter, who started to go, and he's gwyne tost and up to the rack, fodder or no fodder – and d—n it, he can't get no gold by diggin', he's a purty sharp hand at seven-up and poker, and he's got the papers with the spots on 'em.

There goes a motley crew of six b'hoys from St. Louis – veritable rounders and runners "wid der masheen," who are ready for any game from a bruising match to a garroting adventure.

Next follows a tidy little red wagon, drawn by four sleek, well-fed oxen, and followed by a little muley cow, with a merry bell hanging from her neck, of which she seems to be proud, for every now and then she takes occasion to give her head an extra shake just for the music of the thing. The wagon body is compactly stowed with a judicious lot of emigrants' outfitting articles, and in the front, on a calico-cushioned stool, backed up by a pile of clothes and bedding, sits a buxom, blooming young woman, holding up a bright and crowing baby (her first baby – any one can swear to that at a glance), to the admiring view of the stout-limbed and sun-browned young man walking abreast the wheel-oxen – who seems to divide his attention between his team and his load, and cracks his long whip as much for the delight of his baby as for the encouragement of his cattle. They are a cheerful group, and, with hope, health and happiness smiling upon them, no wonder that their hearts keep time to the music on the blue-bird's matin and the lark's brisk orison.

They are scarcely out of sight when slowly creeps along a sort of Jersey cart, drawn by a jaded mule, driven (the driver as usual on foot) by a travel-soiled, care-worn old man, whose stooping figure, furrowed brow, and scant grey locks, speak of many a hard winter and many an unpropitious voyage. . .

Right behind this group – now along side of them – now ahead – is a wheel-borrow or hand-cart chap – going on his own hook, and as independent as a wood-sawyer. He is dressed in a coon-skin cap, red flannel shirt – plentifully besprinkled with buttons and flashy devices of animated creation – with curduroy pants and water-proof boots (with flaming red tops), that come up to his thigh. A leather belt encircles his middle (he has no waist), in which is hung a dirk-knife and a colt's revolver. In his cart is a pair of blankets, a bacon ham, three slices of jerked beef, a few pounds of crackers, and a jug of whiskey, with a few incidental notions.

Now comes two seedy, half-starved, suspicious looking fellows, with nothing but carpet-sacks. They look uneasily about them, as if they fear arrest, cast their eyes up as if to see if there are any telegraphic wires about, dodge into a coffee house for a moment, come out wiping their lips, and start off as if they wanted to put as much space between themselves and the place they recently came from, as possible. . .

TRAINS FOR THE MINES [394]

The travel from this point to the gold mines is still on the increase; it may safely be said not to be scarcely begun yet. The bottom on the Missouri side opposite

[394] *Nebraska Advertiser*, May 5, 1859.

us looks like an army encampment, so many trains are there encamped waiting for grass. Many, however, are leaving daily. We notice trains crossing at this point from various parts of Missouri, Iowa, and Illinois.

Our merchants are doing a heavy business in selling supplies. Such is the run that new supplies are received by them by every boat. If we have such a travel now what may we expect when the season fairly opens.

MILK COWS FOR THE OVERLAND ROUTE [395]

Messrs. Jones & Russell, Thursday, sent out on the express route, eighty milk cows, with all the appliances for an entire dairy at each station along the road. We are told that some of these stations are beautifully located, in spots of choice fertility. Truly, in the case of the express route, cultivation and improvement follows closely upon the footsteps of the pioneer.

PIKE'S PEAK EMIGRANTS RETURNING DESTITUTE [396]

Atchison, May 7.

The Salt Lake Mail has just arrived, and by the courtesy of Mr. TRACEY, the agent of the Hockaday Mail Line, I have just perused a letter from one of their agents at Cottonwood Springs. It gives a doleful and most disheartening account of the Pike's Peak emigration. Large numbers of disappointed gold hunters were already wending their way back to the pale of civilization. But this is not the worst feature of the business. They come back as many of them went, without any means of living on the way. Destitute of provisions or means of conveyance, disappointed and utterly disheartened, with broken hopes and blasted fortunes,

---

[395] *Leavenworth Times,* May 7, 1859.

[396] Correspondence appearing in the *Missouri Republican* of May 11, 1859.

toil-worn, foot-worn, and heart-weary, these wretched adventurers come straggling across the plains in squads of dozens or scores, begging at the stations for goods to eat and a temporary shelter from the driving storms. The well known generosity of the contractors on this line, will doubtless save many a poor fellow from famishing by famine, but what can they do to supply the wants of a starving multitude? Although these men have acted with great indiscretion and improvidence, in their premature and ill-starred journey to the land of golden promise, yet they are fellow-citizens, and the hand of a just and generous government should be stretched out to give them aid in their extremity. As yet no acts of violence have been committed, so far as I can learn, but as the numbers of this crowd of starving wanderers increase, what assurance will there be against scenes of rapine and plunder amongst the trains and stations along the route to Pike's Peak? . . . VIATOR.

RETURNING GOLD HUNTERS [397]
Ft. Kearny, K.T., May 8, 1859

A. K. Miller, Esq: Knowing that you take an interest in all things appertaining to the gold mines, I will let you know how the Pike's Peakers get along out here. They are all returning, and not less than 900 wagons passed here within the last week. They are selling their outfits for almost a song out here... CHARLES D. CURTIS.

EMIGRATION TO GOLD MINES [398]
From our friend, Mr. A. T. Fullerton, who arrived

[397] Correspondence published in the *St. Joseph Journal* May 14, and reprinted in the *Missouri Republican*, May 18, 1859. An article similar in tone appeared in the *Nebraska City News*, May 14, 1859.

[398] *St. Joseph Journal*, May 11; reprinted in *Missouri Republican*, May 14, 1859.

in this city, late Saturday evening from out near Fort Kearny, we learn that the road is lined almost the entire distance with trains of emigrants. For several weeks past, one hundred teams on an average, have crossed the Big Blue per day. There is no suffering on the route, and provisions and provender are as cheap, if not cheaper, and more plentiful than here. All of the hand-cart trains have attached sails to their carts and go "kiting."

POSTOFFICES AT PIKE'S PEAK [399]

We learn that the following postoffices have been established in the new gold mining region, and the gentlemen named appointed postmasters:

San Vrain, Nebraska	C. H. Miles
Lancaster, Nebraska	Charles Blake
Mountana, Kansas . . .	David T. Griffith
Auroria, Kansas	Henry Allen
Coraville, Kansas	Mathias Snyder

It is said that by special contract these offices are to be supplied from Pacific City, Mills co., Iowa. If so, letters and papers should be marked to either of the above, via. Pacific City.

RUSH TO THE PEAK [400]

During the last several days, we have kept a registry of the number of teams that have passed our office, for Pike's Peak, which gives an inkling of the tremendous rush for the mines. Sunday 40 teams; Monday 66 teams; Tuesday 90 teams; Wednesday 108 teams; Thursday 111 teams; Friday 94 teams; Saturday 75 teams; A total of 584 teams.

[399] *Hannibal Messenger*, May 15, 1859.
[400] *Council Bluffs Bugle*, May 18, 1859.

DEPLORABLE ACCOUNTS FROM PIKE'S PEAK [401]

St. Louis, May 18.

The St. Joseph correspondent of the Democrat notices the arrival at that place of one hundred Pike's Peakers, who give deplorable accounts of mining prospects, and the sufferings on the plains. It is estimated that 20,000 men are now on their way thither, all or most of whom are destitute of money and the necessaries of life, and perfectly reckless.

Desperate threats are made of burning Omaha, St. Joseph, Leavenworth, and other towns, in consequence of the deceptions used to induce emigration. Two thousand men are reported fifty miles west of Omaha, in a starving condition. Some of the residents of Plattsmouth have closed up their business and fled, fearing violence at the hands of the enraged emigrants.

PIKE'S PEAK A HUMBUG [402]

People are beginning to make the astounding discovery that "Pike's Peak is a humbug." The fact has been known for a long time, but not till now has it been made evident to the emigrant, bent with feverish haste, to the imagined land of gold. Two or three months ago thousands passed through this state to the "Cherry creek mines," lured thither by the cruelty and atrociously false stories concocted by persons in the border towns, who had outfitting goods to dispose of, and by speculators in the new region, who had town lots to sell. The poor emigrants solemnly believed every word of these fabricated stories. They were determined to be

---

[401] Published in the *Boston Evening Transcript*, May 19, 1859, and in the *New York Tribune* of May 21, 1859.

[402] *St. Louis Evening News*, May 19, 1859. The *Topeka Tribune* of the same date tells of returning gold seekers; while the issue of the previous day speaks of travelers going westward to the purported mines.

deluded – and they have been deluded to a degree that will make their hearts sick and sore for many a long day. . .

When we look upon the whole affair, we wonder how the delusion could have attained such prodigious proportions, and drawn so many hundreds and thousands into its treacherous vortex.

DIRECT FROM THE MINES [403]

The steamer *C. W. Sombart,* came down early yesterday morning from Omaha, bringing Col. J. D. Henderson – better known as Col. Jack Henderson – Capt. Thomas W. Scott, late of the steamer *Twilight,* J. C. Sanders, a mountaineer and government guide of eighteen years standing, and Oscar B. Totten, of St. Louis, who are just in from the Pike's Peak and Cherry creek gold mines, having left Denver City on the 2d inst., and arrived at Omaha on Saturday last, making the trip through in thirteen days. Col. Henderson and party left this city for the mines through a heavy snow storm on the 10th of February last. . .

The first emigrants who arrived this season were handcartmen and footmen, who could easily have got employment had not provisions been so scarce. These became discouraged, returned without prospecting, and of course gave doleful accounts. . .

The col. and his party left Denver City in a four-mule ambulance, met six hundred wagons that would go through, one hundred and fifty more which would go by way of Cherry creek, through Cheyenne pass to California, and passed fully eight hundred, not one-eighth of whom had ever been through to the mines. The most of them turned back at Kearny; a few ven-

[403] *St. Joseph Journal,* May 19; reprinted in the *Missouri Republican,* May 21, 1859.

tured on to Bear [Beaver?] creek, within one hundred miles of the mines, and a still smaller number went within thirty miles.

These gentlemen brought copies of the *Cherry Creek Pioneer* and *Rocky Mountain News,* which were both issued on the 23 ult., and from which we may make copious extracts hereafter.

Col. Henderson and Mr. Totten expect to return, with their families, between the first and tenth of next month.

LEAVENWORTH CORRESPONDENCE [404]

Leavenworth, May 20, 1859.

This day is rendered memorable in the annals of Leavenworth by the arrival of the first overland express in ten days from Denver City. When it is taken into consideration that the great portion of the route traversed, is new and uninhabited, and that the road was first broken by the outward bound vehicles of Messrs. JONES & RUSSELL, the expedition with which the trip has been made is remarkable indeed.

The stages which left this city on the 18th ult. arrived at Denver City on the 7th instant. The entire distance from Leavenworth to Denver City, including all deviations, was indicated by the viometre at 687 miles, but this distance is being shortened by the road agents, and will not exceed 625 miles. The route lies between 39:30 and 40 degrees north. The whole extent, the agents say, is well watered and timbered throughout, (except for 150 miles along the Republican, where the timber is scanty) and is admirably adapted to the requirements of a great national thoroughfare, which it is no doubt destined to become.

The stages just arrived bring four through passen-

[404] *Missouri Republican,* May 25, 1859.

gers and a fraction less than seven hundred dollars in gold dust, some of which is of the flake or scale kind, from Cherry creek, and about an equal proportion coarse, or as it is termed at the diggings "shot gold," from the mountains. . .

[EL PASO MINER] [405]

Mr. Prouty announces in a card that the El Paso *Miner* will not be published, laying the blame on the El Paso Town Company. Better announce its suspension now than after it has been in publication some two or three months.

GOLD HUNTERS ON THE ARKANSAS [406]

By the arrival of Srs. Delgardo, Garcia, and others, merchants of New Mexico, we have received the following interesting statistics. They report having met on the Santa Fe road, between the crossing of Arkansas and Council Grove, the following number of persons, with their wagons and stock, bound for Pike's Peak. The Santa Fe trains met are not included:

Number of men	.	.	.	.	.	.	.	5214
" " women		.	.	.	.	.	.	220
" " wagons		.	.	.	.	.	.	1351
" " oxen		.	.	.	.	.	.	7375
" " horses		.	.	.	.	.	.	632
" " mules		.	.	.	.	.	.	381

These gentlemen also report that the number on this side Council Grove, in their opinion, exceeded those beyond, but they made no memoranda.

FROM THE PLAINS [407]

Messrs. Power, Brown and Cole, three young men

405 *Herald of Freedom*, May 21, 1859.
406 Kansas City *Journal of Commerce*, May 25, 1859.
407 *St. Louis Evening News*, May 26, 1859.

who started from St. Joseph to Pike's Peak, but who turned back 150 miles from Fort Kearny, arrived at St. Joseph last Tuesday. They state that on the 18th inst., a fight took place at the crossing of the Big Blue, between a party of returning emigrants and the ferry-men, in which three of the emigrants were shot, two of them being killed, and the third severely wounded. The names of those who were killed were not given, but it was said that one of them was from Virginia, and the other from Illinois.

FROM CHERRY CREEK TO ST. JOSEPH BY WATER [408]

Messrs. William and Charles Fry, two engineers well known on the Missouri river, passed down, late yesterday evening, on the steamer Sioux City, just from Cherry creek, having come the entire distance by water.

They went out to the mines early this spring, pros-pected all through the richest diggings with old Cali-fornia miners, and were not able to make ten cents per day.

They worked diligently for a period of forty days, and one had forty-five cents worth of gold in a quill and the other three cents worth as the fruits of their labor. They say there is some gold there but in such fine particles that it is absolutely impossible to make wages gathering it with machinery or otherwise.

After having tested the mines fairly, they pronounced them a humbug, built a small flat-boat, eleven feet long by three and a half wide, drawing four inches, and launched it in Cherry creek, came down that stream to Denver City, which place they left on the 4th inst., thence down the South Platte into the main Platte river, to the Missouri river, making the entire distance of eight hundred and fifty miles by water in

408 *St. Joseph Journal*, May 26, 1859.

twenty days. They arrived at the mouth of the Big Platte just as the steamer Sioux City was pushing out Tuesday morning, sold their boat for fifty cents, and took passage on that steamer.

RETURNING PIKERS [409]

An extra train of "Pikers" came in about 2 o'clock yesterday afternoon. – Whew! but warn't they mad?

LETTER FROM THE MINES [410]

Mr. C. L. Cooper, a resident of Washington, Mo., returned from the mines last Tuesday. He met 2000 wagons, 100 hand carts, and 200 footmen on the way out. He reports no suffering either at the mines or on the road. Mr. Cooper has a favorable opinion of the mines, says that he heard of persons making as much as $10 per day, and thinks that any industrious worker can gather $1 to $7 per day.

WELCOME TO JONES & RUSSELL'S EXPRESS [411]

Saturday, last, May 21st, in the goodly year of our Lord, 1859, was essentially a gala day in Leavenworth – an epoch in our history – a day full of hope and golden promise, replete with a thousand joyous episodes and incidents. The sun shone in a sky of unrivalled splendor, and the earth was decked in the fairest of May-day mantles. From noon of day to noon of night there was naught but marching and feasting and enthusiastic acclaims. It seemed as though but one heart beat in our city, and as though one impulse moved all to share in the spirit of the grand occasion.

---

409 *Hannibal Messenger*, May 26, 1859.
410 *Missouri Statesman*, May 27, 1859.
411 *Leavenworth Times* (weekly), May 28, 1859.

The procession began forming about 2 o'clock P.M., at the corner of Main and Shawnee. [A detailed description of the parade follows.]

"PIKE'S PEAK OR BUST" [412]

Among the most significant mottoes we have seen upon any of the wagons going out to the mines, was upon that of an Illinoisan, about three weeks ago, inscribed in flowing letters of red chalk, though not in the most approved style of art – "Pike's Peak or bust." The indefatigable and energetic sucker, returned the other day upon a gaunt and starving mule, that looked as if he had climbed the Peak. He was asked why he didn't go through, "Wal," he said, "he'd got clean on beyond Kearny, and – he busted, so he just rubbed out 'Pike's Peak or bust,' and turned back."

WAGON INSCRIPTIONS – EMPHATIC [413]

We have been somewhat amused in noticing the inscriptions and devices on the wagon covers of the Pike's Peak emigrants. One went through a day or two since, with a large elephant painted over the whole cover. Another had a rude attempt at a pike, with a pyramid to represent the Peak. But the most unequivocal inscription yet, we noticed on the wagon cover of a returned emigrant on Saturday. It reads: "Oh Yes! Pike's Peak in H—l and D—n nation!" We think the man owning that wagon must be of the opinion that he has been badly humbugged, and he thus emphatically expressed himself.

---

[412] *Nebraska City News*, May 28, 1859. This item, showing the first use of the famous expression, was first given me by Dr. R. P. Bieber of Washington University, St. Louis. It was later found among the Willard items.

[413] *Jefferson Inquirer*, May 28, 1859; copied from the *St. Joseph Journal*.

RETURNING EMIGRANTS [414]

The Pike's Peak exodus. The steamers A. B. Chambers and John Werner, from the Missouri river, arrived yesterday, the former bringing three hundred, and the latter two hundred and ten disappointed Pike's Peak emigrants. Many of these are in a state of woeful destitution, and tell the same stories of the hardships and privation as that related by all who have turned their backs on Cherry valley.

[STATEMENT OF OSCAR B. TOTTEN] [415]

. . . Capt. William Parkison of St. Louis, and myself, united in the purchase of an outfit for a six-month's trip, receiving in addition to our party Mr. C. H. Noble, and the two Mr. Wimers of St. Louis, and Mr. A. G. Baker, of Jefferson City. Our party left St. Louis the 24th of September last, arriving at the mouth of Cherry creek the 11th of November; the lateness of the season making it necessary to prepare for winter quarters we could not then prosecute our search for gold. We found some sixty men at said place, with four houses already up and many more in progress; these men assured us that they found their mining prospects flattering. We erected our houses, which required much hard labor, after which we prospected at such times as the season would permit. On the Platte and tributaries we found from one to five cents in the pan, and with water at hand could have made from $3 to $5 per day; but water being so difficult of access we deferred mining till we could get into the mountains. About the 20th of March, Col. Henderson, W. H. Wignall, G. W. Wainwright and myself started. We penetrated about sixteen miles into the mountains, it

---

414 *Missouri Republican,* May 28, 1859.

415 Totten's letter to the *Missouri Democrat* and published in that paper on May 30, 1859.

snowing and storming badly we were obliged to come out – but good color was found. We returned to Cherry creek, took provisions, and started for Boulder, where we met J. Ely, Judge Tounsley, and Williamson, and at their solicitation we accompanied them to the north ford of St. Vrain's fork of the Platte – distance fifty-five miles northwest of Cherry creek and twenty-five miles from Boulder. We found the snow in the gulches from three to five feet in depth; we shoveled off the snow, built fires to thaw the ground, and dug down eighteen feet, finding the gold to pay from one to six cents to the pan from within two feet of the surface down the whole distance. Our party immediately struck off claims and commenced getting out lumber for sluices, and making all preparations for mining. As the weather was unsettled and stormy, and being short of provisions, Col. W. H. Wignall and myself returned for supplies to send them. I also saw some fine looking quartz ledges (and having mined two years in California consider myself a judge), and so well satisfied was our whole party that we found gold in paying quantities, that Col. H. and myself have returned for our families. Among those who have a party there is Mr. B. F. Langley, of California, who has mined for years in that country, and is perfectly satisfied with the prospect that he got at this place. I do not make this statement to induce any person to go. I advise no one to go. I make these assertions, which time will prove correct, for the purpose of showing the nature of this stampede, and freely assert that those who have returned with such alarming accounts of no gold there, give nothing from actual experience, for every reasonable man knows that they had not had time to develop the mines. OSCAR B. TOTTEN, St. Louis.

RELIEF BY GOVERNMENT FOR THE DUPES OF THE
PIKE'S PEAK FRAUD [416]

It has been suggested at Washington, that supplies
of provisions should be immediately forwarded for
the relief of the poor fellows who have been entrapped
by what is now so commonly considered the Pike's
Peak humbug. The following is from the Washington
correspondence of the *Philadelphia Gazette:*

"The accounts from the returning adventurers, who
were misled by the fabricated representations of gold
at Pike's Peak, are awful, and doubtless will grow
worse, unless some prompt relief should be extended
to the starving and suffering bands. . .

Something efficient ought to be done, and the sense
of the country would at once respond to whatever assistance the president might assume to render, in the
absence of any appropriation or provision for such
an exigency. Congress would not and dare not refuse
to cover any exercise of discretion thus humanely employed, . . No words of resentment are strong enough
to express the indignation which should be visited
upon the authors of all this misery, who, in their greed
of gain, sent to the east such wilful exaggerations of
the wealth of that barren and desolate region."

The correspondent of the *Gazette* is in part correct
when he brands much of the representations from
Pike's Peak as a swindle. Return[ed] travelers and
letter-writers from the spot, who have grossly lied
about it, deserve no quarter. With all this, however,
much of the misrepresentation has been made in ig-
norance. There is no doubt gold diffused throughout
that region and many who perceived that fact, but not
sensible of how much it would cost to get it, really
believed that fortunes might be made out of it. How

[416] *Missouri Republican*, May 31, 1859.

far journals at the "starting" points and "outfitting" points are morally responsible for the mischief that has been done, the editors themselves can perhaps best tell. They know best whether when they were puffing the mines, they did so with a full belief that their puffs were warranted by facts, or wilfully exaggerated or circulated what they knew either to be untrue or did not know to be true, for the sake of drawing trade and travel through their "points." These remarks apply also to some of the journals of St. Louis, one or more of which have until a very recent date given prominence to glowing descriptions of the mining prospects.

Returning Miners [417]

A steamboat from the Missouri river arrived yesterday with two hundred and fifty returned Pike's Peak gold seekers. Their accounts of the new land of Ophir are by no means encouraging. One of the passengers says his party, consisting of fifteen men, worked the first day at Cherry creek with all the freshness of an ambitious zeal to make a fortune, and their combined efforts resulted in the gathering of about thirty cents worth of the "shining thing," – or two cents apiece. Everybody is leaving Pike's Peak, except a few land speculators, still intent on realizing something from their investments at Auraria and Denver City.

Retreat of the Ten Thousand [418]

For the last two weeks the streets of this city have been filled with emigrants returning from Pike's Peak. They all appear jaded and care-worn, and all complain of the criminal duplicity of parties, who, as they al-

[417] *Ibid.*, May 31, 1859.

[418] *Atchison Union*, June 4, 1859. The *Herald of Freedom* of June 4, says that the Jones and Russell coaches which left Leavenworth May 30 carried 2156 letters to the gold region.

lege, to promote certain local and other interests, have involved thousands, and tens of thousands in irretrieveable ruin. We have conversed with many of them, and each tells the same tale of deceit, disappointment and wretchedness. Some have been through to the charmed mountain, others turned back before reaching it, and all unite in denouncing it as we are inclined to think it really is, one of the hugest humbugs that has been inflated since the great South Sea Bubble. Most of the victims are farmers and mechanics from Michigan, Missouri, Ohio, Indiana, and Illinois, and appear to be a temperate and well informed people. The lesson has been an impressive one – we trust it will be a warning to them, and to others hereafter.

THE EBB OF THE GOLD TIDE [419]

We are informed by Col. Jacob Hall, of Independence, that the mail conductor on the Santa Fe route, who arrived on Saturday, reports having passed 2500 wagons on their return from Pike's Peak. The advance guard is beginning to arrive already. While this army is passing off down the river, it would be well for our city authorities to put a police force on duty, as there may be occasion to protect property from the depredations of any lawless characters that may be with the trains. Citizens should also be exceedingly careful to guard their houses against robberies, for our resident thieves will take advantage of the rush of gold hunters to carry on their work on a large scale, thinking the blame will all be saddled on the Peakers. Too much vigilance cannot be exercised the next two weeks.

THE HAND CART [420]

Two men arrived yesterday who have hauled a hand

---

[419] Kansas City *Journal of Commerce,* June 7, 1859.
[420] *Ibid.,* June 8, 1859.

cart all the way to Fountain City and back. They were from Chicago, and left here about the 25th of March, with 600 pounds of provisions, tools, etc., traveled all the way there, worked in the South Park mines nine days, and have returned well and hearty, with a small amount of gold, and considerable experience in traveling on the plains. They were offered seventy-five cents for the cart but did not sell it.

PIKE'S PEAK BAND [421]

Our city has been enlivened for the last few days by the superior music discoursed by the brass band, from Freeport, Illinois, now enroute for Pike's Peak.

That "music hath charms" is fully demonstrated from the attention given by the hundreds of listeners to the superior performance of this band. May they succeed in their new enterprise, and when Gabriel blows his trumpet, be found in tune.

HUMBUG COMPLETELY EXPLODED [422]

. . . This has been, not inaptly styled, the age of humbugs, and Pike's Peak has indeed proved the humbug of humbugs. It has no parallel in the history of civilized nations, unless we accept the famous "South Sea Bubble.". . The spectacle of 100,000 people simultaneously abandoning all the comforts, conveniences and endearments of home, and setting out, many of them on foot, and without a dollar in their pockets, and with barely provisions to last them a week, upon a journey from five hundred to a thousand miles, over a wild and inhospitable region, all animated and almost run-mad with gold greed; and then, after a lapse of a few weeks, coming back, begging, starving, cursing,

[421] *Council Bluffs Bugle,* June 8, 1859.
[422] Editorial in the *Hannibal Messenger,* June 9, 1859.

and many of them hopelessly ruined, is one never before witnessed, and one that teaches such a lesson as, in our opinion, will prevent repetition of a similar act of folly for a long time to come.

PIKE'S PEAK – GOLD IN ABUNDANCE [423]

Leavenworth, June 10.

The express coaches have arrived with advices from Denver City to the 1st. They contain extraordinary reports, calculated to renew the gold fever. Rich nuggets had been discovered at the north fork of Vasque's creek, principally coarse gold, and decomposed quartz. Great excitement prevailed all through the country, and the statements of the yield of Gregory's and other mining companies, were almost fabulous. A company from Indiana is making from $150 to $200 per day.

There are other accounts of the same character, not varying in the least. Many of the letters received seem to have been written under the prevailing excitement, and are in strong contrast to previous gloomy reports.

The dust taken out by the Gregory Mining Company alone is estimated at $20,000.

Denver City was almost depopulated by people leaving for the mountains.

Provisions continued scarce.

Twenty thousand dollars had been offered for claims.

Subsequent arrivals will determine for certainty the truth of this information, but none of the accounts now differ in tenor. Rich specimens of gold were received by this arrival, and this community are intensely excited by the intelligence.

[423] *Boston Evening Transcript,* June 11, 1859. This was news of the Gregory lode discovery near present Central City. This was the first lode gold found in the region and the first real basis for mining development.

PLATTE RIVER NAVIGABLE [424]

We hear every few days of small craft boats making trips from Auraria to this place in fifteen and twenty days. In the course of time larger boats will try it and make the trip.

FROM INDEPENDENCE [425]

Independence, June 11, 1859.

. . . Pike's Peakers are still returning in great numbers, and some venturing out, not satisfied until they see for themselves. The reports received by express at Leavenworth, so encouraging, are not believed here by those who have returned. . .

[EXPRESS FROM THE MINES] [426]

Leavenworth, June 13.

The Pike's Peak Express arrived here yesterday, seven days from Denver City, bringing $1,400 in gold dust.

Accounts continue to be received of the same nature as by previous arrivals, and the practicability of the mines is considered as completely established.

The first supply trains had arrived, and provisions met with a steady sale at good prices.

PIKE'S PEAK PANIC! [427]

We were at Council Grove, one day last week, where we went on purpose to ascertain, if possible, the facts in reference to the return of the Pike's Peak emigration. There is among returning emigrants, even, so many conflicting stories, that it is almost impossible

[424] *Nebraska City News,* June 11, 1859.
[425] Published in the *Missouri Republican,* June 15, 1859.
[426] *Nebraska Advertiser,* June 16, 1859.
[427] *Kansas Press,* June 13, 1859.

to learn anything reliable. One thing is certain, however, the Pike's Peak emigrants are returning by thousands. In fact, a perfect panic has seized the whole emigration; whether with or without cause, the future alone must develop.

As we approached Council Grove, from the bluff, this side for miles, each side of the Grove, we were in full view of the Great Santa Fe road, and it was literally lined with returning Pike's Peakers. Hurrying into town, we found it jammed full of men, women, and children; within one hour we counted over one hundred wagons on their return. Never did our heart bleed, for a set of people, as it did for these men and women. A month ago, with joyful countenances, and bright hopes for the future, they left all, and set out for this "modern Ophir!" But now, with dejected countenances, some of them the very picture of despair, are returning to their homes, one half of whom are ruined men.

We talked with many of them, some (but a very few) claimed to have been through to Cherry creek, some had been to Bent's Fort, some to the Arkansas, some to the crossing of the Cottonwood only, whilst many turn back at the Grove. We were told that nine hundred wagons turned back in one day at Bent's Fort. Twelve hundred and fifty wagons started down the Platte in one train, from Cherry creek. . . Mutterings, loud and deep, were heard against those who had humbugged them, as they said; men were swearing like pirates, who, from their awkward manner, in using the profane, had evidently been pious in the states. We should not wonder if we heard of terrible outbreaks, when this emigration reaches the river towns, murders and robberies may become common. Great suffering, it was said, existed on the plains; men who had given out, were, in many instances, left to starve. The wheel borrow, hand cart, and foot emigrant, were literally starving.

Two hundred wagons, it was said, after turning, and traveling two hundred miles on the back track, at the crossing of Cottonwood, had heard more favorable reports, wheeled around, and are now enroute for the mines. . .

UNPARALLELED RICHNESS [428]
Gold! Gold!! Gold!!!

Times office, 2 P.M., June 20.

A few hours since we published a large extra *Times,* containing a most satisfactory statement from Horace Greeley, A. D. Richardson and Henry Villard (Times correspondent) in reference to the immense gold deposits at the Gregory diggings, and the general richness of the whole gold district. . .[429]

FROM THE MINES [430]

. . . In coming down the river from Omaha last week, we fell in company, at the mouth of Platte, with W. G. Wignall, N. L. Bolton, G. W. Wainwright and D. C. McCarron who had just reached the Missouri from Denver. They came all the way down the Platte in a small boat of their own construction, having made the trip in thirteen days. They report recent encouraging discoveries; they together with two others formed a company of six who took out in one day $96 in shot gold. The reason they leave now is the scarcity of provisions. . .

---

[428] *Leavenworth Times,* June 20, 1859.

[429] This famous statement, as published for the first time, in the *Rocky Mountain News* (Denver), will be given below, on pages 376-82.

[430] *Nebraska Advertiser,* June 23, 1859.

Pike's Peak [431]

The Pike's Peak fever has broke out again, and is raging to an alarming extent in and about the city of Leavenworth. It has already carried off several enterprising citizens of that place, and more are preparing to follow in the same direction. To be candid, we more than half believe that there is gold in these famous mountains in remunerative quantities. We regarded the whole thing as a huge humbug until we read Greeley's statement (if it is his), which we publish in another column, and to which we direct the attention of our readers. We never admired him as politician but as a business man, Greeley's opinions are as good as the gold. Time will determine whether Mr. Greeley really wrote the statement to which his name is signed, and if he did, whether his opinion was correct. We can well afford to await development of events.

[431] *Atchison Union,* June 25, 1859.

LETTERS FROM THE MINES
APRIL TO JUNE, 1859

LETTERS FROM THE MINES
APRIL TO JUNE, 1859

[G. N. WOODWARD] [432]

Auraria, April 3d, 1859.

. . . When I last wrote you, I was on the Arkansas river and expected to stay there through the summer, but have changed my mind and am now at this place, on the South Platte, and shall probably remain here most of the summer, but cannot tell where or what we shall do in this wild distant land. This place and Denver City, are all one, that is, side by side – they are quite large places, mostly built of hewed logs. Within the last few days there are some good frame buildings gone up. We have six stores; about 12 or 15 saloons; one hotel, etc. etc. There is more drinking and gambling here in one day than in Kansas City in six – in fact about one-half of the population do nothing else but drink whiskey and play cards.

I have written twice before to you not to come to Pike's Peak this spring, for the reason that there is no diggings as yet discovered that will pay as much per day as Henry can make in the office. . .

The report now is that they have found gold in great abundance in the South Park, in the mountains, but no specimens have yet reached here. . .

Denver City, or Auraria, K.T., April 9, 1859.

I have waited for some days and Smith's express has arrived at last, thank God.

---

432 Kansas City *Journal of Commerce*, May 12, 1859.

I saw men from the mountains yesterday and I can depend on their statements, and they have found coarse gold in abundance from $12 to $30 diggings per day, this changes the whole matter and if you wish to come in the spring I have no objections, in fact, aside from the hard journey I should be glad to have you here. . . Yours truly, G. N. WOODWARD.

[A. C. SMITH] [433]
Denver City, April 8th, 1859.

. . . Since I have been here I have been round considerable, and believe that I have had pretty nearly as good an opportunity as any one to judge of the truth or falsity of the many reports that we heard upon leaving the states. And, in the first place, my impression is an extremely favorable one, and each succeeding day only increases my faith in the richness of the mineral deposits. . .

Judge Lynch, to day, hung a man for murder, and as the corpse of the murderer lay side by side with that of the victim, no one could but feel that the act justified the result. The particulars are as follows: Some three months since a party, consisting of an old man by the name of Beingraff, in company with his sons, Antoine and Philip, son-in-law, Jno. Stuffle, and John Ellis, all Germans, arrived here from Louisville, Ky. Yesterday the body of Antoine Beingraff was discovered in Clear creek, about seven miles from this place. His corpse was brought to town, and I was one of the coroner's jury. During the examination evidence was brought to light that tended to throw suspicion upon the son-in-law, Jno. Stuffle.

---

[433] *Leavenworth Times,* of May 4; reprinted in the *Missouri Republican,* May 8, 1859.

Parties were sent out in pursuit of him, who captured him this morning about daylight, as he was returning to his home. He was brought to town, and court was convened, Judge Smith presiding. A jury, composed of twelve of the most respectable persons in our community, were selected, and the prisoner was given a fair trial. The jury, after an hour's deliberation, returned a verdict of wilful murder. He was committed, without bail, to stand his trial at the first term of the criminal court, and the prisoner was remanded to the custody of the sheriff. One of the officers of Judge Lynch's court then moved that the prisoner be allowed one half hour's time to prepare for eternity. He was allowed a spiritual adviser (Rev. Mr. Fishe[r]), and at the expiration of three quarters of an hour was led to the place of execution. The rope had been thrown over the limb of a tree, and a wagon placed under it. The prisoner came very composedly forward, mounted the wagon and made a speech in which blasphemy and raving took a prominent part. The wagon was drawn from under him during his speech, and he died with scarcely a struggle. The old man is also missing, and it is supposed to have been a premeditated plot between the son Philip and son-in-law to get the old man and Antoine out of the way, and take the property, as Antoine's body was robbed when found. Both of the other Germans are now in custody, and it is thought that the brother will swing to-morrow. Gen. Larimer conducted the defence, and Henry Allen, Esq., the prosecution.

Capt. Preston, who went ahead as an *avant courier* of Russell & Jones' express, arrived yesterday, and returns tomorrow. His mission was to select the shortest route and arrange stations for the ensuing season. Capt.

Wm. Smith also arrived to-day, just as the execution took place. As time and opportunity allow, I will endeavor to keep you somewhat posted in regard to matters in this neck of timber. Ever yours, A. C. SMITH.

(P.S. Enclosed in Mr. Smith's letter was a small lot of gold dust, of the scale kind, remitted as a subscription to the *Times* for one year).

LIFE IN DENVER CITY [434]

Denver City, K.T., April 17, 1859.

After 44 days of terrible struggling, I have at last reached my journey's end.

Denver City is a log city, containing about 100 cabins. Corner lots range in price from $50 to $500. Inhabitants are Indians, Mexicans, and white people – about equally divided – all hard cases. Drinking and fighting all the while; some one killed nearly every week – now and then one hung.

There is a great deal of humbug about the gold stories you see published in the newspapers. The best mines yet found will only pay about three dollars per day, when worked by old miners, with plenty of water. . .

At present there is very little here to induce people to come out, yet they keep coming. A great many are also going home, disgusted with the country. . .

If you know of anyone who has the gold fever very bad, you will do him a kindness, perhaps, by telling him to wait awhile – there is no hurry. If there are any paying mines here, there will be just as many another year.

---

[434] This is an unsigned letter written by a young man to his father in Rochester, New York. It was published in the *Rochester Democrat* and reprinted in the *New York Tribune,* May 28, 1859.

[OLIVER CASE] [435]

Denver City, April 19th, 1859.

I am now living in Denver City, six miles from the old place Montana which has gone the way of all flesh. We have not done any mining this winter on account of the weather, and because there have been no mines discovered that will pay over from one to two dollars a day, without quicksilver, for the gold is so light that it all runs off with the sand. We have made enough to keep us all summer in provisions.

We expect to go into South Park next week, as the news from that locality have been very flattering, and that from persons who have given up this country as a humbug and had resolved to leave for California if they were not successful in South Park. If the news turns out true it will make El Paso the best point in this country, and I shall try to get an interest in it. They have also started a town in the South Park called Park City, and I will write from there if I stay there any length of time.

At the town of Arapaho, fifteen miles from here, they have been at work all winter, and have not made money enough to buy food. They could make from $10 to $15 per day with a sluice if they could only save it, but the gold is so fine, that it rises to the top of the black sand and the water washes it off. . .

The emigrants are beginning to come in and it is a sight to see them. Some come with oxen and wagons and some with mules, but the most have come bringing hand carts, and when they have got here they have no provisions, no money and no clothes, and can make nothing here at mining. The report here is that the people on the road out have got the names of all the

[435] Kansas City *Journal of Commerce,* June 7, 1859.

letter writers from this country, and if not true they will hang the letter writers and burn the towns. They will have a good time. . .

The Arkansas road is the best leading to this country in every respect. All persons that have come the Smoky Hill route say it is very rough and no trail half the way. The report is here that twelve or fourteen persons have frozen and starved to death.

Some of the boys that were in our train went into the mountains a few days since prospecting, yesterday they came in after provisions, and they say that they can make from eight to fifteen dollars per day in the South Park. I shall start on Monday and shall report in a few days.

I do not think this country will pay as an agricultural district at all. The soil is too sandy, and the seasons too dry and cold. Yesterday it snowed all day and froze very hard, but the snow is going faster to-day than it came.

Give my respects to all the boys and tell them to stay where they are until they hear something more reliable than has been published. Very affectionately, OLIVER CASE.

[J. T. PARKINSON] [436]

Denver City, April 20, 1859.

J. Jewett Wilcox, Esq. – Sir I am much pained to have to inform you of the sudden and violent death of our mutual friend Captain P. T. Bassett, by the hands of John Scudder. The cold-blooded and dastardly assassination caused much excitement here, and if he, the murderer, had not effected his escape, he would have expiated his crime on the gallows. Bassett was entirely unarmed, and without warning was cooly shot

[436] *Missouri Democrat,* May 21, 1859.

down by the coward, the ball taking effect in the right lung, and after lingering in great agony all day, he died about 7 P.M. This was on the 16th. I have been appointed his administrator and will attend to his affairs hereafter. After I have ascertained what they are, I will communicate with you that you may let his mother and daughter know of them. You will please write to his relatives with whom you are acquainted and break to them the sad news.

The inquest proved clearly that Scudder came to Bassett's house, called him out and asked him if he said "that he was a thief!" Bassett instantly said, "no." Scudder then asked him if he told the mail carrier not to deliver his (Bassett's) letters to him for fear of their being opened. Bassett answered "yes." Scudder at once drew his revolver and shot him.

You know the description of Scudder, and I wish you to advertise a reward of $500 (for his apprehension and detention) in the Missouri *Republican* and *Democrat,* the *Valley Tan* at Salt Lake and in some paper in San Francisco. Scudder escaped on Sunday night, while in the hands of the sheriff, the day before Bassett was buried. Yours truly, J. T. TORKINSON [Parkinson].

[A. CUTLER] [437]

EL PASO, April 20th, 1859.

E. S. LOWMAN, ESQ – DEAR SIR – I arrived here a few days ago, and after having visited all the other towns in this portion of Kansas, I must say that in my opinion this has the finest site and the best prospects for a large town, of any of them. It is situated at the mouth of the canon leading into the South Park, about

[437] *Herald of Freedom,* May 28, 1859. Cutler was the chief promoter of the town of El Paso.

five miles from the base of Pike's Peak, at the terminus of the Smoky Hill route, and on the best and safest road from the Arkansas to Cherry creek, as by this road the divide is but about four miles wide, while by Jim's camp [438] it is more than fifty. I saw some gentlemen to-day who had just returned from a tour in the Park. [439] They showed me a specimen of shot gold they brought out with them. They say that they made five cents to the pan from the surface down, and had not got to the bed rock yet, but had sunk one hole ten feet deep. They say the weather is so cold in the mountains now that they cannot work, but are going to return after they put up some buildings here. There are a large number of persons about to build here, I understand, and I should not be surprised to see this place soon outstrip Auraria and Denver. A large number of persons are leaving Auraria and coming over here for the purpose of erecting buildings, believing that this place is bound to be the metropolis of this country. I am told that all of the buildings are to be of a good class, being constructed either of boards or hewn logs, and none of them to be less than eighteen by twenty in the clear and one and a half stories high.

I camped a few days at the Boiling springs. [440] They are indeed a great curiosity. The water has very much the taste and appearance of the Congress (Saratoga) water. We made some bread, using the water of the spring, without any soda, and we made better light bread than we could make with soda or saleratus. Saleratus sells here for 75cts per pound, so you may judge of the amount saved in using this water for making bread alone. . .

---

[438] For data on Jimmy's camp see the preceding volume in this series, page 104.

[439] South Park.

[440] Manitou springs, near present Colorado Springs.

[HENRY ALLEN] [441]

Auraria, K.T., April 24, 1859.

Editors Council Bluffs Bugle: You will before this reaches the Bluffs, see some of those who have returned from this country discouraged, and will no doubt hear all kinds of stories about the mines; people, etc. Some that have come out here say that they did so on the strength of a letter that I wrote to my family, dated December 19, 1858, and they blame me. Well, all I have to say is simply this. All that I wrote is strictly true, and was at the time.

The first companies that came through this spring, were hand-cart trains. They were out of provisions, foot-sore and out of spirits. As soon as they arrived, they wanted work. We had no extra tools, nor provisions to give them. We had been here all winter among the Indians, that we had to feed, rather than have a difficulty with them; consequently were nearly out of provisions ourselves when the first train arrived. The most of these men started back without even prospecting, and reported along the road that there was no gold here, and that we were starving.

The next trains were horse and mule trains. They came in after meeting those that were returning, and not seeing plenty of gold in our streets, and large stores of provisions, their minds being already prejudiced by those they met on the road, readily concluded that there was nothing here, and some after staying one day, some two days, and some sooner, started back for the states. This is about the truth of the whole matter. . .

Now Messrs. Editors, for the gold news. I am running one sluice on Dry creek, which is paying very

[441] *Council Bluffs Bugle,* May 18, 1859. In introducing the letter the *Bugle* says that Allen had been reported killed for writing favorable letters about the mines.

well, and tomorrow I shall send up two more. These diggings pay from $3 to $5 per day to the hand, clear of expenses. On Clear creek, they make about the same. On Cherry creek, near Russelville, there is a company mining and doing well, in fact better than we are doing on Dry creek. On the Platte they are making from $3 to $8 per day, according to the manner they are working.

You will probably see some that were here. If you do they will tell you that they could get the "color" anywhere, and those that went with me, will tell you that I always got from two to ten cents to the pan, but the dirt was to carry from one hundred to one hundred and fifty feet. They complained that there was too much stripping or that the water was too far off, and so they took the back track. . .

With this letter I send you a newspaper,[442] in which you see that Mr. Langly, Henderson, Palmer Goodwin and others have just returned from the mountains, and have brought seven or eight ounces of coarse gold and auriferous quartz. Mr. Langly has been in the mountains about eight miles from here the most of the winter, and you may depend upon what he says being true. He is an old California miner – he reports the snow so deep in the gulches they cannot be worked to advantage for 40 or 50 days. I have examined the gold brought in by him – it is coarse, rough and altogether different from that we get in the mines here. I need not say any more about the mountains as you will see the news from them in the paper sent you.

Now Messrs. Editors I wish before closing, to say that I do not desire to persuade any one to come here, but if you have any persons in Council Bluffs who wish

---

[442] This was the first issue of the *Rocky Mountain News,* published April 23, 1859.

to make money, and are willing to work for it, they can do so by coming here and bringing tools with them and going to work; but if they calculate to make a fortune by coming out here and drinking whiskey, or loafing, then let them stay where they are – this is no place for them.

There is a great deal of suffering on the Smoky Hill route – a company arrived last night. The Indians stole their cattle – they packed a part of the way and thirty hours before they came in they eat the last of their pony.

Another company of ten on that route have buried three of their number, and I think will soon bury two more. They report no water for one hundred miles, only that found in buffalo holes. Yours truly, HENRY ALLEN.

[CLARENDON DAVISSON] [443]

Arapahoe, K.T., foot of the Rocky mtns., May 2, 1859.

I have been here with the Chicago company for a fortnight. The company consists of the following:

Messrs. J. P. Horton, David P. Foote, S. B. Sampson, George N. Simmons, E. G. Reynolds, A. C. Kent and John F. Greene.

These gentlemen are the most substantial and best miners here. Their first attempts on this stream proved failures, but their discovery of coarser gold at a point 25 miles above this, in the mountains (or nearly 40 miles N.W. of Denver at the mouth of Cherry creek) on Vasque's fork of the Platte is likely to turn out better. They are packing up as I write to start today to give it a trial. Probably a hundred others will follow or go with them. It is but a short distance from

---

[443] This letter was published in the *Chicago Press and Tribune,* June 3, 1859. Davisson had been the commercial reporter on this paper, according to editorial comment preceding the letter.

Long's Peak on a small stream which they have called Chicago creek,[444] which a branch of this will furnish them water. Pine in abundance surrounds the locality. A flock of 150 big horns, or mountain sheep feed near it, which with black tail, antelope, elk and bear, will furnish most of their food. Foote shot a large gray wolfe and a small martin the other day. Another of the party shot an antelope or gazelle with a revolver this morning on a plateaux near their camp.

I am told that three more Chicagoans, Messrs. George Star, Rufus Cook and Clark Morton arrived in Denver yesterday and will probably go to the mountains. E. A. Bowen, W. H. Valentine and L. Wiley from La Salle, Ill., are still here. I was elected a delegate to and drew up the preamble and resolutions for a first convention for the purpose of organizing a state.[445] (Signed) CLARENDON DAVISSON.

[JOHN M. FOX] [446]
Office of Leavenworth City and Pike's Peak
Express Co. Denver City, K.T., May 8, 1859.
JOHN S. JONES, Esq. – Dear Sir: Our tedious march

---

[444] Chicago creek enters Vasquez fork, or Clear creek, at present Idaho Springs. George A. Jackson had discovered gold here in early January 1859. See "George A. Jackson's Diary, 1858-1859," in *Colorado Magazine*, XII, 201-214. The *Chicago Press and Tribune* of June 6, 1859 prints J. P. Horton's letter of May 12, which speaks thus of the Chicago creek diggings:

"We are starting back tomorrow with all our teams, tools and two months provisions. It is about 26 miles from this place and about forty from Cherry creek. We can get our wagons within about eight miles of the diggings and pack our animals the rest of the way. The diggings are situated on a small stream in the mountains, which, by the way, we have named Chicago creek and which empties into a stream known as Vasque's fork of the South Platte. About a half mile this side of this stream is another of the same size with five or six beautiful pure soda springs on its banks. It is beautiful for raising bread."

[445] The "state of Jefferson." See L. R. Hafen, "Steps to Statehood in Colorado," in *Colorado Magazine*, III, 99-100.

[446] *Daily Missouri Republican*, May 25, 1859. Fox was Denver agent of the Leavenworth and Pike's Peak Express Company.

is at last ended, and we are now safely located in Denver City, the much talked of golden city of the mountains. Our progress from Junction to this place was necessarily slow, inasmuch as we had to open a new road through a country, about which but little was known – to contend with the severity of the weather, the fatigue and complete exhaustion of many of the mules, and the many obstacles incident to an enterprise of this magnitude. I can truly say, sir, that I believe our road is the best, in all respects, that can possibly be made from Leavenworth City to the mines. Wood and water are in abundance over the entire route, excepting about one hundred and fifty miles upon the Republican, where there is some scarcity of timber – in fact a great scarcity for emigrants; but our stations can be rapidly supplied from the pineries lying some thirty miles distant from Cherry creek.

Nearly all the station-keepers, men and employees upon the road, express themselves satisfied with this location. Some one or two swear they will not stay, Murphy (at no. 19) among them.

We have had two desertions only. Our nearest station to this place is forty-three miles. An intermediate station must of necessity be made until Mr. Williams returns and shortens the road, which he expects to do, saving a distance of fifty miles or over.

Much of the country over which we passed is eminently adapted for agricultural pursuits, and a great deal of it almost or wholly worthless.

Permit me to say, that I think Col. Preston [447] missed the *chute* both in going back and coming out, being too far north on his outward trip, and a great deal too far south when he returned.

[447] Preston had gone out ahead of the stagecoaches to select a route for the express company.

I would be more explicit in regard to the exact line we traveled, but Mr. Williams will return in a short time, and can better point it out on the map. . .

We reached Denver City on yesterday (Saturday), May 7th. Gen. Larimer received us, and has treated us with extreme courtesy and hospitality. . . The people were much gratified at our arrival. . .

It is true that there has been a large emigration here within the last month; and also true that nearly all have gone back to the states. Their report cannot be relied on from this fact, if no other; a large majority of the emigrants were men who lacked energy and industry, with no means – almost starving – finding no provisions in the country (the Mexican supplies not having yet arrived) – seeing no gold lying upon the ground – discovering Cherry creek to possess properties similar to other waters, and not one bit yellow – wanting the "vim" to prospect and prosecute what they had undertaken to do – became dissatisfied, discouraged, furious and raving mad; took down the Platte – some in boats, some on foot, and some with their teams – turning back all who were on their way here that would listen to their tales.

That there is gold here, the dust which I send with this letter is sufficient evidence. As to the quantity, no man can form any idea. I have had conversations with a good many persons, and have met some who say there is no gold here. Others say that all that is necessary to develop the country and open out and lay bare the rich treasures of the Platte, Cherry creek, Plum creek and the gulches of the mountains, is for men of untiring energy and perseverance, with means to keep them until they can open a claim, to come here and go to work, and work faithfully.

One of the passengers in the stage went up the Platte

six miles, washed out two pans of dirt and got 25 cents, and I saw the money! He is an old California miner, and says he is perfectly satisfied.

I heard this evening that rich quartz discoveries had been made near Boulder Town, some thirty miles up the mountains. One of the specimens which I send – $58.75 – is the result of 14 days labor of one man with a common cradle, within three miles of this place, handling his dirt four different times.

There is a large plateau of land lying between Cherry creek and the Platte, to work which successfully, water is absolutely necessary. To obtain a sufficient quantity, it will be necessary to ditch and turn Plum creek. A company is being formed for this purpose. Out of this the proprietors expect to reap colossal fortunes. It is my candid opinion that it will take some time to find out what is here. I heard that one man would not take $10,000 for his claim. I do not know that is true. I will write you again and again as I make discoveries. . .

The Indians stole in this vicinity last night, one hundred head of horses and cattle. None of ours. Old Raven, chief of the Arrapahoes, came in to day, and will have a grand "pow-wow" on to-morrow.

I will remark that but very little mining has been done. The spring is backward, and the snow long in leaving the mountains. Yours, respectfully, Jo. M. Fox.

[CLARENDON DAVISSON?] [448]

Denver City, May 9, 1859

Saturday evening, last, 7th inst., the first and long looked for mail and passenger stage of the Leavenworth and Pike's Peak Company arrived here. It was out twenty days from Leavenworth, but now that the way stations are formed, and bridges made, it is thought

[448] *Missouri Democrat,* May 25, 1859.

fourteen to fifteen days will suffice to make the trip. The new route thus laid out via Republican fork of Kansas river, seems to be a good one, according to the report of the stage company, though about the same length as the old Ft. Kearny and Arkansas routes, hilly and sandy on this end, and destitute of timber for fuel, for an equal distance with the others, or some 100 to 160 miles. I think it quite probable this express line will do a good thing in opening up this region of the far west, but from present appearances however, the company owning it will not enrich themselves, at least not in the legitimate way of carrying mail matter and passengers.

We have had sickening rumors for a week past of horrible sufferings from freezing, starvation and robbery of emigrants by way of the Smoky Hill route. . . The number of arrivals here to-day and for the past few days shows an increase; but the tide seems to be equally strong in the homeward bound direction. Most, or very many of those now returning, go by skiffs, dugouts, etc., upon the Platte, the water of this river being now in a good navigable state, down stream, for such craft. I have a flat-boat built, and propose launching it in the morning, and may beat the stage, which carries this letter, as the current is about six to eight miles an hour.[449] It may be carried to the bottom, however, as it is somewhat dangerous. About fifty boats of one kind and another have sailed in the past week, with an average of four to five persons in each. It is estimated that 2500 persons have left for home in the last twenty days. This is in addition to those that turned back be-

---

449 Davisson's venture on the Platte was unsuccessful and he turned to the stagecoach. He arrived at Leavenworth on June 3, according to the *Leavenworth Times* of June 4, 1859.

fore reaching here, induced by the intelligence received from others on their return. Probably as many more reached here in that time; but I think there are no more people in the whole mining country to-day than there were on my arrival here a month ago. The towns (or villages which they all are, and small ones too) seem quite deserted by whites, and as many Indians as pale faces are now in and about this village. Yesterday a dozen or more Indian braves, of the Arapahoe tribe, rode in, accompanied by their head chief, [Little] Raven,[450] and two of his brothers. They collected at the office of Judge Smith, to hold a talk with the whites, relative to an arrangement with the government for the sale of their territory, and some aid to their nation to become an agricultural and a civilized one. They are already about as civilized as the whites here, and far more temperate; indeed, I've not seen an intoxicated Indian in the mining region. The interpreter, one Antoine Du Bray, was drunk at the meeting, and falling off his seat as he sat down by the intelligent looking and dignified Quaker-like chief, Raven, rendered any attempt at a pow-wow a farce; and it was immediately adjourned for a day until the interpreter should become sober. I may yet be enabled to give you an account of the talk.

As for gold news, or reports from the mines, nothing reliable is heard; but many rumors of paying diggings are afloat, which are not credited by Yours truly, D.

[J. S. W.] [451]

Denver City, May 12, 1859.

I arrived in this goodly city on the third, by way of

450 Little Raven was a prominent Arapaho chief.
451 *Leavenworth Weekly Times*, May 28, 1859.

the Smoky Hill route. Have been fifty-nine days on the journey, two weeks of which time I have lived on cactus and wild onions, occasionally killing a crow and other birds.

Many who left your thriving city a few weeks ago, are now sleeping the last silent sleep of death.

For the benefit of the traveling public, I would say that the route by way of the Republican fork, established by Jones, Russell & Company, is the best and shortest route yet found to this land of gold, . . .

I find that a great many emigrants have already arrived and returned to the states discouraged, never having struck a pick in the earth. Most of those remaining are glad to be rid of them. Large numbers are gone to the mountains to prospect; but the mountains being yet covered with snow render it difficult for much mining to be done. . .

This place is graced by the presence of some ten fair ladies, among whom I find several from your city — among others, Mrs. Wade, Mrs. McLaughlin, Stoddard and Hall. . .

Rumors are afloat that Jones, Russell & Company, are to be interested in Denver City and make it headquarters in this country. The enterprising Messrs. Parkinson of St. Louis, are opening a large farm below the city and will doubtless make a good thing of it, as vegetables will be very high. . .

Since writing this, good news has arrived from the mountains, about twenty miles from here: A man brought down some hundred and fifty dollars, the result of one week's labor.

You will hear from me regularly hereafter. Yours, J. S. W.

[HENRY ALLEN] [452]

Auraria, K.T., May 13, 1859.

Editors of the Bugle: Since I last wrote to you, I have been in the mountains prospecting, and have found what I have heretofore hoped for – mountain gold.

On the second of this month I started with Wm. M. Slaughter, formerly of Plattsmouth, N.T., for the mountains, and after traveling over a very rough and mountainous country for thirty-five miles, came to what is called Vasquers fork, or Clear creek. We had just arrived when an old-fashioned snow storm – such as are only gotten up in the Rocky mountains, commenced. After the storm we commenced prospecting. We had nothing but a pick and pan, to work with.

We commenced with the top dirt, and in every pan got from two to ten cents worth of dust, and sometimes more – no float gold; but all rough gold. I send you a specimen panned out by myself.

We found plenty of quartz. I prospected over two miles on the main creek and one mile on a small branch, and found gold about the same all the way. We sunk a hole about 80 yards from the creek, on a high bank, to the depth of 13 feet, and found boulders. As soon as we got to the boulders, we found coarse gold. There are now about 150 men at work on that creek and are making from $3 to $12 per day to the man, and are working under great disadvantage.

It is impossible to get in sluices until we can whip-saw the lumber, as we cannot cross the mountains with wagons, nor can we pack rockers or sluices to the diggings until we can find a better road in.

The Cherry creek and Platte ditches are progressing

---

[452] *Council Bluffs Bugle,* June 8, 1859.

slowly for want of provisions. The Georgia company have finished a small ditch on the Platte river and are doing tolerable well. You will see from the *Rocky Mountain News* what they make; but you must recollect that these diggings are not in the mountains, and the gold all scale or float gold.

W. N. Byers, editor of the R. M. News and myself will start on tomorrow for the mountains, and when we get back you will hear from me again.

We intend to go into the mountains as far as we can get for the snow. I am confident we will find larger gold – some pieces that we brought out this time weighed 95 cents, some 50 cents, and so on down.

There has been a regular stampede on the Platte route back to the states, by some hundreds who have never been within fifty miles of this place. Others came and stayed one or two days – swore at everybody and then left; and others came in – begged something to eat, stole a mule or pony and left, telling on their way back, all kinds of hard tales. Some of them went so far as to bury D. C. Oaks in effigy, and then told he was dead. Others reported that Gen. Larimer and Allen had left the mines through fear of being mobbed; but they are all here and will be here digging gold when those that went back will wish they never had started to the gold mines of Kansas.

Since writing the above, S. B. Kirkland, of St. John, Harrison county, Iowa, has come in from the mountains. He has found quartz gold. I have examined it and am certain that it has just been taken from the vein and has not been carried far from the place where it was found. I shall visit that place on my trip, certain.

You may depend upon what I write being true. Yours truly, HENRY ALLEN.

[RUFUS CABLE] [453]
DENVER CITY, May 14th, 1859.

I wrote you about two weeks ago, and was then starting for the South Park – and have just returned. It proved a failure. The rivers there are frozen 4 and 5 feet, with no indication that summer ever makes its appearance in that quarter. The air is so light that one cannot breathe freely. Located at the base of the "Snowy range," and some 5,000 feet [454] above the level of the ocean, it is not very probable that any one will ever attempt to work there.

Coarse gold has been found, but not in sufficient quantities to justify working. I saw, last evening, some very good specimens of gold taken out 25 miles from here, and will start for that place the day after tomorrow. We will very likely stop there and work until October, when I go home, for I would not spend another winter here for any amount of money. Those sentimentalists, who sigh for a "lodge in some vast wilderness," can find several by coming to this place.

I suppose by the time you receive this 'twill be pretty generally known in the states that Pike's Peak is a humbug. Every day crowds are starting for home, cursing the country and those who wrote letters from here last fall. . .

There are now some twelve or fourteen hundred men out here; a great many of them have started within a day or two to the mines, which I think will pay, at least enough to justify out here.

There is a great deal of sickness here at present, confined principally to those who came here last fall; as

453 *Western Weekly Argus,* June 4, 1859.

454 The elevation of South Park is over 9000 feet. Gold was discovered there later in 1859, and in 1860 the population in the park was 11,603, according to the federal census.

the breaking up of the winter has commenced, they have the pleurisy and some fever, but generally speaking, I think the country, remarkably healthy. I have not been ill an hour since I left home.

The next letter you get from me I will be able to inform you whether or no the mines will pay. If they do I will be home next fall; if not, ho! for Arizona or Sonora. Yours truly, RUFUS CABLE.

[LEAVENWORTH TIMES CORRESPONDENT] [455]

Denver City, May 15, 1859.

Odd as the situations have been into which I was brought in the pursuit of my correspondential career, in divers quarters of the globe, I must yet confess that I never perpetrated a letter under more peculiar circumstances than surround me at the moment of this writing. I find myself in the town of Auraria, seated at a rudely-constructed cotton-wood table, in the office of the *Rocky Mountain News,* the first paper ever published in the western part of the territory. From the interior of the locality a full aspect of the firmament is impeded by a few shingles only, that represent the roofing, and furnish an altogether insufficient protection to the inmates from the ill favors of the weather. A few feet from me, one of the proprietors of the establishment is busily engaged in kneading dough and frying bacon. On my right the stalwart form of an Arrapahoe warrior, wrapped up in an immense buffalo robe, obstructs my vision. His coverless head is adorned with the half-grown antlers of a young elk. His eyes watch attentively the every motion of my pen, and express wonder and amazement at its doings. Two squaws – the chattels of the red-skinned gorgo-

---

[455] *Leavenworth Times,* June 4, 1859. The correspondent was most probably Henry Villard.

nizer – are spread on the floor with several pappooses in nature's unmodified dress, all chuckling over and diving into a cup full of sugar given to them by a good-natured typo. Every once in a while the aborigines interrupted the run of my thoughts, addressing me in the soundless but telling language of signs, which were to convey propositions to trade for such articles of civilized luxury as sugar, coffee, and the dearly beloved whiskey, etc., and the reader will, therefore, know to explain the imperfections of this communication.

The working material of the *Rocky Mountain News* arrived here some six weeks ago, and three numbers of the paper have already appeared; but thus far the proprietors have found very little encouragement for the continuation of their enterprise. The publication of a newspaper in this section of the country is evidently premature. The population is floating, and have hardly means enough to satisfy physical wants. Promises by the owners of town property, given in consideration of divers puffs, constitute about all the inducements that have been held out to the publishers. Their main income is derived, for the time being, from the sale of single copies, at 25 cents, to gold seekers. One or two numbers of a proposed rival weekly have also made their appearance in Denver City, but, Mr. Merrick,[456] (formerly of Elwood, K.T.) the proprietor of the concern, (the same as was reported as having been killed by another printer while crossing the plains), very wisely sold out, a few days ago, to the publishers of the *Rocky Mountain News,* and struck for the mountain diggings.

To-day the Holy Sabbath was diversified by the convening of Judge Linch's court, for the purpose of try-

---

[456] But one issue of Merrick's *Cherry Creek Pioneer* was published.

ing one of the many horse and cattle thieves that infest
this country. A few days ago a number of horses and
mules – among the latter, one belonging to the express
company – were found missing. It was soon ascertained
that they had been abducted by some of the above-
mentioned gentry. A number of men started in pursuit, and last night one of the depredators, that had
been caught some ten miles above here, in company
with animals, was brought into town. A preliminary
meeting of the citizens was called, which resulted in
the determination to have the offender tried by the
people at large, at nine o'clock this morning. At the
appointed hour the interior of the Denver House – a
hotel without floor, ceiling, windows or partitions –
was crowded with a motley crowd, and appeared to
be anxious to vindicate justice according to the simple
but effective law of the border country. The variegated
costume, armament and bearing of those present gave
the scene an unusual interest to my eyes. Mexicans,
mountaineers, Indians, gold-hunters and other elements
of population were mixed together in picturesque con-
fusion.

The meeting was called to order; a president and
secretary elected, and, upon motion, a body of twelve
selected from the multitude. The prisoner was then
brought in; counsel procured for him, and the trial
proceeded in regular style. The evidence was heard,
as well as the arguments for the defense and prosecution; the case summed up by the president of the meeting, and finally given to the jury. The evidence not
proving the commission of the theft by the defendant,
he was acquitted with a warning to leave the neighborhood at the earliest possible moment.

There appear to be a good many lawless fellows
here, but also a strong body of good men, who seem to

be determined to suppress depredations upon the person and property of themselves and others at all hazards. The lynch law will undoubtedly be frequently resorted to during the coming summer.

About two weeks ago the Apaches stole about 150 head of cattle from this vicinity. At that time thousands of Indians were encamped all along Cherry creek. They disappeared, however, on a sudden, and since then no more Indian outrages have been heard of. They were Utes, Cheyennes, Apaches, Comanches, and Arrapahoes. Of the latter a few are still camping about here. Little Raven, the chief of their nation, has had a "talk" with the people of this town, and pledged his word for the preservation of peace and law and order by his people.

This morning quite a number of emigrants arrived here from the states. Several mule trains from western Missouri and eastern Kansas were among them. I conversed with a number and found most of them full of confidence, in spite of the stories related to them by disappointed gold-seekers. A good [many] miners came in also from the mountains. I saw gold dust to the amount of several hundred dollars in the possession of eight of them. The greatest difficulty most of these men had to encounter was the want of money necessary to buy a few week's provisions before making for the mountains. They mostly started with a few pounds of flour and meat, which was, of course, exhausted in a few days, and compelled them to return with what little dust they had found to replenish their stock.

Second Letter –   Denver City, May 18, 1859

The arrivals from the states have been very large during the last two days. On the day before yesterday, Capt. Russell's trains, consisting of a long array of mules and ox-teams, and about 150 men. Also, a number of trains from Nebraska, Iowa, eastern Kansas, and Missouri.

During the course of yesterday eleven trains came in, one of which from DeKalb county, Illinois, was the largest. It numbered 57 men and some 20 heavy wagons. I estimate the aggregate number of arrivals for the last 48 hours to be over 1200, the great majority of which at once struck for the mountain diggings. But two or three representatives of the fair sex were among the crowd.

Most of the trains came in on the Platte route. They report that the arrivals would have been five thousand, had it not been for the panic that was created on the route by the ridiculous stories of some disappointed, weak-minded Pike's Peakers, all of which are without the least foundation of truth.

I learn that the panic has also spread along the Smoky Hill and Arkansas route, and that thousands are returning without ever having seen the "land of Israel."

The Regulators, of whose expedition after a den of horse thieves I spoke in my last, have returned. It is reported that they hung their informant for misleading them.

A man by the name of Thorpe came in here from Boulder City at a late hour last evening. He has with him one pound and four ounces of shot gold, the result of his own labor. His treasure has been shown around generally, so that there can be no doubt as to the reality of the thing.

I shall embark this morning upon a tour of inspection through the various mountain diggings, in order to enable myself to give you some reliable information as to the inducements that are found there by miners.

[HICKORY ROGERS] [457]

Denver City, May 20, 1859.

Friend Oscar [Totten]: I have delayed writing to you for some time, in order that when I write I might give you reliable information. Since you left there have been some good prospecting done in the mountains, and all are now satisfied that this will prove a good country. Small parties are daily coming in from the Jackson,[458] Jefferson [459] and Boulder diggings with gold, and all agree stating that the mines pay from five to ten dollars per day. Since you left we have a daily express to this point from Leavenworth City. The emigration is still coming in large numbers, and very few are returning. Town property is on the rise, and buildings are fast being completed, and new ones erected.

I have not heard from Baber since you left, but Mr. Williams, of Boulder, was here a few days since, and says that Baber has a good thing and is doing very well. G. Russell and party arrived a few days since, and have located on the Platte, at or near Dry creek, and are doing a good business, making about ten dollars to the man per day.

Judge Smith left here a few days ago for the Clear creek mines, in company with Mr. Allen. Mr. Allen, who has been in the mines, stated to me that his party had averaged from five to ten dollars per day while at work.

Provisions are very scarce and high, and unless we

---

457 *Missouri Democrat,* June 9, 1859. Rogers was one of the officers of the Denver City Company.

458 The Jackson diggings were at present Idaho Springs.

459 The Jefferson diggings were on the upper waters of South Boulder creek. An unsigned letter written from Boulder, May 17, 1859, and published in the *Jefferson Inquirer* of May 28, 1859, says: "We have laid out a town, 25 miles from this, in the mountains, and called it Jefferson City. It bids fair to be the largest town in the mines. I am a stockholder. The best diggings in the country have been discovered out there."

have some trains from Missouri I am afraid we will
suffer. Mexico is not able to support our increasing
wants in this way. We have secured the express in Denver, and they are now making the trip through in from
twelve to fourteen days, and they speak very highly of
the road up the Republican fork. We are anxiously
waiting for Wynkoop, when a drawing of our town
will take place. I will see to your interest in all things
until your arrival. Write and let me know when you
expect to start. Several new stores have started since you
left, and "Zavon's Lightwig" [Taos Lightning?] has
gone up the creek. Old Bourbon and Kentucky have
taken its place. Everything looks bright and prosperous
once more. If you can conveniently bring a good soft
wool hat, I would like for you to do so – size, 7½, wide
rim. I will write again in a few days. In haste, yours,
etc., HICKORY ROGERS.

[LEAVENWORTH TIMES CORRESPONDENT][460]
Denver City, May 22, 1859.
Since my last, no further news of importance from
the mountain diggings has reached this place. The rush
in that direction, however, does still continue increasing in volume, in consequence of the numerous arrivals
from the states during the last two days. Among the latter were two large trains from Kansas City, consisting
of over thirty wagons, propelled by magnificent horses,
who had stood the trip in a remarkable manner. They
came by way of the Santa Fe route. Trains from Iowa
and eastern Missouri also crossed Cherry creek and
South Platte on yesterday, on their way to the mines.
Things again look dreary. There is a perfect stagna-

---

[460] *Leavenworth Times,* June 4, 1859. Apparently this was written by
Henry Villard.

CROSSING THE PLATTE

From a contemporary sketch of 1859

tion in this place and Auraria. Provisions are nearly exhausted and no other trading is done. . .

The South Platte is still rising, in consequence of the continued melting of mountain snows. Its present depth and swiftness renders fording impossible and the use of a ferry necessary. This institution is about the only money-making concern in both places. Its charges for carrying across are shamefully enormous, although it is itself but a rickety fabric, whose original cost did not exceed $300, and effects a trip twice in five minutes. Foot passengers are charged 25 cents each way; a man and a horse, 75 cents; a wagon and two animals, $1.50 and $2.00 in proportion. The general strike for the mountains to reach which the Platte had to be crossed, causes the receipts some days to amount to several hundred dollars. . .

LATER. – Several parties have just arrived from the north fork of the Clear creek (a tributary of the Platte). They report the discovery of rich quartz veins in the valley of that stream.[461] I shall start myself tomorrow for that locality, to investigate the matter in person. Ex-governor Beall, of Wisconsin, who has been around here for some time, is also embarking upon a mining expedition.

The fourth train of coaches arrived yesterday noon from Leavenworth City, in twelve days and six hours. They brought Mr. Martin Field, who at once took charge of the mail department of the express company, with his usual vigor.

A dinner party was given on yesterday, at the office of the express company, to Little Raven, the chief of the Arrapahoes, and four of his most distinguished warriors. Little Raven is a very sensible and friendly dis-

---

461 The Gregory lode and other gold veins discovered in this Central City region went far to save the situation from collapse.

posed man. He handles knife and fork and smokes his cigars like a white man. He was so well pleased with the entertainment that he invited Dr. Fox to come to his lodge and sleep with him, which invitation was, however, not as heartily received as given. A band of Navajo Indians are also encamped between the two places.

In the evening of the day before yesterday Dr. Greenly, a native of Chioga county, Ohio, but lately of Iowa Point, died from the effects of bilious fever. He was a dentist, and came here last fall. He was universally respected, and had met with considerable success in his profession while here. To die in this half-savage country, and among people whose only thoughts are bent upon the rapid accumulation of wealth is certainly a hard lot. The unfortunate man was decently interred last evening.

[J. M. Fox][462]

Denver City, May 31, 1859.

To John S. Jones, Esq. Dear Sir:—Unlike my former letters, I hope this will be eminently gratifying to you. Reports from the mountains, come in daily, and all without a single exception, confirm each other, and are encouraging in the highest degree, and beyond the possibility of a doubt true... Mr. Gregory, the discoverer of these leads, was entitled to one claim by the working, and one extra claim for the discovery he made. He sold these two claims for $27,000. Bates was offered $20,000 for his but declined. One man sold his claim to the Indiana company for $50.00. In a few days afterwards they sold it for $4000.

These things are true! You can state them anywhere

and everywhere without fear of their being successfully refuted. The mines are but scarcely opened. It will take the whole summer to develop them. . .

The utmost excitement prevails among the miners as well as the citizens of Denver. Every one is anxious for a claim. Nearly all have gone to Gregory. Fields states that there is at least already in the mines, and between this place and there, from 2,000 to 3,000 people. . .

The Gregory diggings are not by any means the only ones which are paying. The large piece of gold which I send was taken out at Jackson's mines – miners there are making from $5 to $8 per day to the hand. This specimen will give some idea of the character of the gold taken out at these diggings. Its weight $4.20. . .

A gentleman here has kindly promised me $10,000 in gold, which I will certainly take, if I can buy that am't of gold dust, trusting however to receive that amount by return stage. I know it would be of the greatest advantage to receive that amount of gold from here.

You can set down the unparalleled richness of this country as a fixed fact. Old California and Georgia miners say that the quartz leads are the richest they have ever seen.

You may be certain that I will commence remitting to you when the money begins to come in. I have some $2000 or $3000 on hand, but have been anxious to get as much as possible to buy dust, thinking that would meet your wishes best.

The quartz of which I will speak below, is exceedingly hard, and will have to be blasted. I intend making some arrangements to supply trading posts in the mountains, also to send expresses once a week with mail to the mines.

The two pieces of quartz rock are specimens of what

is found in large quantities. This kind of rock cannot be worked successfully without a crushing mill. With a magnifying glass, you can discover small particles of gold. These two pieces are supposed to be worth 50 cents. Yours respectfully, J. M. Fox.

[MAT. RIDDLEBARGER][463]
Denver City, June 1st, 1859.

. . . For the past week the reports from the mines have been of the most exciting and cheering kind – for instance, about six days ago Jim Winchester arrived from the mountains and reported that he saw five men take out near two thousand dollars in one day. This started nearly everybody in town for the mountains, and in two days nearly everybody living here broke for the diggings. To-day I was informed by reliable (so I consider them) men that, on last Saturday, three men in one day, made $260, and that five men made $590. And this is vouched for by several men who came down from the mountains. Still, again, claims have been sold for from two thousand to eleven thousand dollars. One company from Chicago sold two claims for $21,000, and yet the returning crowd will swear there is no gold in the country. No, and if it was left for them to find it, never be developed. I am satisfied, and I care not how many comes or goes. Mark my words, the gold is here, and will be found. Yours, MAT. RIDDLEBARGER.

[MISSOURI DEMOCRAT CORRESPONDENT][464]
Denver City, June 3, 1859.

The excitement about the Gregory diggings is still

---

[463] Letter written to Riddlebarger's sister, Mrs. William G. Barkley, and published in the Kansas City *Journal of Commerce,* June 14, 1859. Riddlebarger was a newspaper publisher in Colorado in 1860-1861. See D. W. Working, "Some Forgotten Pioneer Newspapers," in *Colorado Magazine,* IV, 99.

[464] *Missouri Democrat,* June 15, 1859.

on the increase. Authentic information has been received here to-day of the striking of still richer leads by a prospecting party conducted by Mr. Gregory, and the sale of the latter's claim on the original Gregory lead for not less than twenty-one thousand dollars, to a party of four, of whom three, viz: Amos Gridley, E. B. Henderson, from Cass county, Indiana, and a Wm. Allen, from Fulton county, Illinois, are known. The brothers Defrees, who also had two claims on the Gregory lead, sold their interest for $7,500, and $7,000. One of the Ziegler brothers sold a claim, which he had bought a week previous for $50 for $6,000. A number of other sales of less magnitude are also reported. Marshal Cook, of Doniphan, bought a claim of Gregory for $600, and Samuel J. Jonce, a well known railroad contractor, two of another individual, for three mules and $600 cash, respectively.

The exodus to the diggings is daily extending in volume. Denver City is all but deserted. I do not think that more than three hundred are now living within the city limits.

Mr. St. James, a Mexican trader of Scotch descent, now domiciled in this place, bought this morning something over two ounces of nugget gold consisting of lumps weighing from two to seven pennyweights each. They had been brought in from the Jackson diggings.

The arrivals from the states still continue to be large, but returning emigrants have become a scarce article. Whoever lands here, now at once steers for the mountains. . .

The first supply train sent out by Messrs. Russell & Jones, consisting of twenty-five wagons drawn by six splendid mules each, and loaded principally with groceries, arrived a few minutes ago. It is a real godsend in view of the general scarcity of almost all articles of trade in this place. The animals look as sleek as though they had just left Leavenworth City.

[HENRY ALLEN][465]

Auraria, June 4, 1859

Dear Bugle: – Since I last wrote you, I have again been in the mountains, and to give you all the news of that trip, would probably subject you to some considerable leisure, as you have some in Council Bluffs who have been to the gold mines, "seen the elephant," and returned to tell all about the Rocky mountain gold field – the burnt cities – the men lynched, etc. etc., and are now telling yarns they do not believe themselves. I am very sorry for some of them, as I know they have made big calculations from the reports they have received here, and would have done well if they could only have been prevailed upon to stay. No doubt I have received my share of curses, from some, for they have done it here before they left! Well, here is the news since I last wrote; and what I write you, you may depend upon, as I have been to the diggings myself, and worked them, and have now three companies there at work. The editor of the *Rocky Mountain News,* and myself, went prospecting, and for the result of the trip, I refer you to his and Mr. Gibson's report in the *Rocky Mountain News.* Since then I have been sick.

Foster, Slaughter, and Shanley, are in the mountains and are making money – not by the dollar, but by the hundreds of dollars! The diggings pay from three to ten dollars to the pan. This is certain! It is quartz diggings, and there is no knowing how extensive they are. One thing is certain, I have traced some of the leads nine miles and there is plenty of them. I have found them from twenty-five yards to thirty rods apart, and

---

[465] *Council Bluffs Bugle* (extra), June 22, 1859.

running parallel for miles, and all pay about the same. Now you can tell whether there is any gold in the country, or not, and how near the truth we have been writing to the *Bugle*. I always told you that what I wrote, should be the truth, and I have never wrote you one word respecting the gold mines, that I have to take back. Mr. Kinesman, through whose kindness you will receive this, will give you his experience in the mines – and you may relie on what he says, for he was in the mines, and that is more than a great many who have gone back with their big yarns about there being nothing here. I can hear nothing of my family, I fear they have gone back with the stampede – though I hope not.

The constitutional convention for the formation of a new state, met on Monday. I will post you of its progress.

Health is very good. Provisions tolerably plenty just now; flour, $15 per hundred. Yours truly, HENRY ALLEN.

DENVER CITY CORRESPONDENCE [466]

Denver City, June 4, 1859.

. . . For the first two weeks after the opening of the express office in this place, it occupied a log cabin of a rather primeval description. A few days ago, however, the headquarters were removed, to a more civilized abode, consisting of frame, and affording a plentiful supply of light, of which the former windowless haunt had been entirely destitute.

The express company carries, as you are undoubtedly already aware, the United States mail, and their mail department is a branch of their business, of great importance, extent and profit. It is under the superintendency of Mr. Martin Field, formerly of the St. Louis,

[466] *Missouri Republican,* June 15, 1859.

and lately of the Leavenworth City post office. Although but recently arrived, he has already succeeded in systematizing the discharge of his onerous duties, and his office now presents that perfect mechanism that alone is apt to secure satisfaction to the public in mail matters.

The post office is, of course, a place of general rendezvous, crowds of emigrants and immigrants, diggers, traders, mountaineers, etc., can always be seen in and about it, retelling their hopes and disappointments. . .

The flow of humanity from and to this point is still continuing. A marked diminution in the number of arrivals has, however, taken place within the last few days. The thousands that took the back-track after a short trial of this country, have evidently succeeded in turning the masses they met on the road. From a reliable source I learn that more than eight thousand people are now moving eastward on the Platte route alone. I do not think that the arrivals for the last ten days exceeded a thousand souls. . .

Notwithstanding the gold discoveries in the mountains, to which I propose to allude further below, general prostration still prevails here, in every branch of human activity. The merchants do nothing worth mentioning. The whiskey-sellers can afford to lie flat on their bellies, and real estate operators enjoy a desperate calm in their respective lines of business. The sheriff, a few days ago, sold sixteen lots and a good log house, all located in Auraria, for $50.00. An individual bothered me almost to death with his attempts to saddle some of his "choice corner lots" on my shoulders. For $35 I could have had six lots on the principal street of Denver City, with one good and one half-finished log house. But I respectfully declined the investment. . .

Several days ago we were startled by reports of several strikes of vastly rich leads of quartz gold in the north fork of Clear creek or Vasquez creek, a tributary from the northwest to the South Platte. I immediately set out to sound the matter, and found the news for once to be based on reality. A Mr. Gregory, of Georgia, and several parties from Indiana, Illinois, and Iowa, while prospecting the hilly banks of the above mentioned stream, had come across a number of streaks of burnt quartz, which induced them to use their picks and shovels and pans forthwith. The first pan of surface dirt yielded over four dollars worth of particles of gold of the finest description. . .

The excitement about the quartz leads is intense. On my way back to this point, I met at least two thousand individuals bound in their direction. Denver City and Auraria now look as dull as New England villages on Sabbath day. Everybody joins the general rush. . .

FROM A PRIVATE LETTER [467]

Denver City, June 4, 1859

Well, Dear C, I have, as yet, been unable to write to you and fulfill the contract we entered into. For weeks after my arrival I was greatly discouraged – so much so, that I was several times on the eve of returning to Leavenworth.

But now – just hold my hat! Did you speak of gold – why, bless your soul, old TIMES, there's nothing but gold here. I have just returned from the mountains for provisions. For the last nine days the miners have been averaging over $75 a day. My claim is a little outside of the Gregory diggings, where the fellows are making great strikes. In truth the mines have turned out richer than we ever believed they could. California can't hold a candle to Pike's Peak. . .

[467] Unsigned letter published in the *Leavenworth Times* of June 18, 1859.

[FREDERICK KERSHAW][468]

Denver City, June 10th, 1859.
Dear Brother, I arrived here on Monday the 6th, and started immediately for the mountains, some forty-five miles from this town, in company with Horace Greely, Mr. [B. D.] Williams, Mr. [Henry] Villard, Mr. [A. D.] Richardson, and Mr. Martin Fields, post master. . .

We visited some ten different mines that were in operation and got a report from each, of the amount of gold taken out. I think I can safely say, that four thousand dollars is now being taken out per day by them all. . . FREDERICK KERSHAW.

[HENRY VILLARD][469]

Denver City, June 10, 1859.
About an hour has elapsed since I returned from my second excursion into the mountains, in company with Horace Greeley. This trip I did not make under as favorable circumstances as the first one. I met with a variety of accidents, among which was a severe sprain of my left wrist, in consequence of my being thrown and dragged by the mule I rode, is not the least. Worn out as I am bodily, I yet propose to give you a brief account of the incidents and results of this journey.

We started early on Tuesday morning, and after having crossed the Platte on the Auroria ferry, reached the foot of the mountains in less than two hours. Mr. Williams, the superintendent of the express company, had

---

[468] *Hannibal Messenger,* June 26, 1859. Mr. Kershaw was from New York, according to Henry Villard, in his *The Past and Present of the Pike's Peak Gold Regions* (edition of 1932, edited by L. R. Hafen), 51.

[469] Published in the *Leavenworth Times* of June 20, and reprinted in the *Missouri Republican,* and in the *Missouri Democrat* of June 24, 1859. The letter is unsigned, but from its content one can learn that it was written by Henry Villard. Accounts of this trip to the diggings may be found also in Horace Greeley, *An Overland Journey,* etc. (1860), 115-127; A. D. Richardson, *Beyond the Mississippi* (1867), 177-203.

very kindly placed one of the express company's coaches at the disposal of the excursion party, and accompanied it personally, so as to secure the greatest possible speed and comfort to all participants.

From the river crossing up to the base of the Table mountains and through the beautiful valley intervening between the latter and the first range of the Park mountains, we found the road, and adjoining natural meadows, literally covered with trains of gold seekers, herds, camps, ranches, etc., all of which had sprung up only during the last eight days.

Both banks of Clear creek, which we reached after fifteen miles' travel, we found lined with hundreds of wagons and tents, and thousands of grazing animals. Among the inhabitants of these extemporized canvass towns, a number of fair ones in "bloomer" figured most conspicuously.

The news of Mr. Greeley's proposed visit to the diggings had out-traveled us, and when we reached the high bank of the creek, a large crowd greeted the arrival of the distinguished editor with a chorus of hearty cheers.

Clear creek, naturally of extreme swiftness, turbulence and depth, had experienced a considerable increase of its natural mightiness, in consequence of heavy rains in the mountains, and crossing with the coaches became, therefore, entirely impracticable. We accordingly saddled our mules and plunged simultaneously into the creek.

Old Horace, although seriously hurt below his left knee, by the accident mentioned in my last letter, never faltered, but spurred his mule and made against the current. As the water reached our animals' bellies, and finally rose up to our knees and thighs, and Horace yet stood the ground, the spectators could not help bursting into another enthusiastic cheering, which was kept up until the entire party had safely landed on the opposite side...

From the ascent of the first mountain, which can boast of the pleasant feature of an all but perpendicular height of at least sixteen hundred feet, to the diggings, the road consists of ups and downs of great steepness. The ravines – they hardly deserved the names of mountains – separating the several mountains, while the commanding peaks themselves were covered with heavy quartz and granite boulders.

Mr. Greeley did not find the saddle as comfortable as the editorial tripod. He soon experienced the fatigue and soreness of green equestrians, and he greeted the camping hour at noon and in the evening with unmistakable satisfaction. His aforementioned sufferings to the contrary nothwithstanding, he proved himself a very congenial traveling companion, whose inexhaustible stock of good humor and wit tended to alleviate the hardships of the trip to a large extent.

The distance from Denver to the diggings being over forty miles, and the road rugged to an extraordinary degree, we did not arrive in the valley of the Rallston fork of Clear creek, along which Gregory's diggings are located,[470] until nine o'clock next morning.

I merely wish to add, that since my first visit at least fifteen more sluices have been completed, and twenty more paying leads struck, along which hundreds of claims have been taken. The majority of the latter are, however, not yet worked, on account of the want of blasting means. I estimate the quantity of gold turned out to be at least $3,500 per day.

Mr. Greeley expressed the firm belief that in two

---

[470] Gregory diggings were on a branch of North Clear creek, not on Ralston creek.

months there would be some five hundred sluices in running order, and by dint of which the yield of dust was calculated to reach $150,000 per week. In the evening Mr. Greeley addressed an assemblage of some three thousand miners – about half the entire number now working and prospecting in Rallston valley. He spoke on the gold resources of the country, giving his opinion as to their origin and quantitative and qualitative extent, advocated the speedy organization of a community of a state, and admonished his hearers to be temperate in their habits of life, spurn groggeries and gambling hells, and wound up by wishing all them the very best success in their mining operations. He was frequently interrupted by enthusiastic applause.

Mr. Williams, the superintendent of the express company, succeeded him in some eloquent and logical remarks, in the course of which he took occasion to refer to the willingness of the company he represented to facilitate the course of the miners with the states at the lowest possible rates. He explained the arrangements made by the company for the shipment of dust, transportation of mails, etc., all of which was received with evident gratification by the audience. A. D. Richardson, Esq., of the Boston *Journal,* and Judge Smith then followed in a few humorous remarks, after the delivery of which the meeting dissolved. A short time before dinner the following day, we took the back track and reached this place without any accident happening to any one, except the writer of these lines, in the already mentioned manner.

Mr. Greeley will spend eight or ten days in this vicinity.

The constitutional convention, after perfecting a permanent organization, and appointing committees on the different heads of the convention, adjourned on the morning we left for the mountains, 7th inst., until the first Monday in August.

The arrivals from the states have averaged several hundred per day for the last seventy-two hours.

[THE GREELEY REPORT][471]

Gregory's diggings, near Clear creek,
in the Rocky mountains,
June 9th, 1859.

The undersigned, none of them miners, nor directly interested in mining, but now here for the express purpose of ascertaining and setting forth the truth with regard to a subject of deep and general interest, as to which the widest and wildest diversity of assertion and opinion is known to exist, unite in the following statement:

We have this day personally visited nearly all the mines or claims already opened in this valley, (that of a little stream running into Clear creek at this point;) have witnessed the operation of digging, transporting, and washing the vein-stone, (a partially decomposed, or rotten quartz, running in regular veins from south-west to north-east, between shattered walls of an impure granite,) have seen the gold plainly visible in the riffles of nearly every sluice, and in nearly every pan of the rotton quartz washed in our presence; have seen gold (but rarely) visible to the naked eye, in pieces of the quartz not yet fully decomposed, and have obtained from the few who have already sluices in operation accounts of their several products, as follows:

Zeigler, Spain & Co., (from South Bend, Ind.,) have run a sluice, with some interruptions, for the last three weeks; they are four in company, with one hired man.

---

[471] This statement was first published in an extra of the *Rocky Mountain News* on June 11, 1859, and subsequently was widely printed in the east.

They have taken out a little over three thousand penny-weights of gold, estimated by them as worth at least $3,000; their first day's work produced $21; their highest was $495.

Sopris, Henderson & Co., (from Farmington, Indiana,) have run their sluice six days in all with four men – one to dig, one to carry, and two to wash: four days last week produced $607: Monday of this week $280; no further reported. They have just put in a second sluice, which only began to run this morning.

Foote & Simmons, (from Chicago) : one sluice, run four days: two former days produced $40; two latter promised us, but not received.

Defrees & Co., (from South Bend, Ind.,) have run a small sluice eight days, with the following results: first day $66; second day $80; third day $95; fourth day $305; (the four following days were promised us, but, by accident, failed to be received.) Have just sold half their claim, (a full claim is 50 feet by 100,) for $2500.

Shears & Co., (from Fort Calhoun, Nebraska,) have run one sluice two hours the first, (part of a) day; produced $30; second, (first full) day, $343; third (today) $510; all taken within three feet of the surface; vein a foot wide on the surface; widened to eighteen inches at a depth of three feet.

Brown & Co., (from De Kalb co., Ind.,) have been one week on their claim; carry their dirt half a mile; have worked their sluice a day and a half; produced $260; have taken out quartz specimens containing from 50 cents to $13 each in gold; vein from 8 to 10 feet wide.

Casto, Kendall & Co., (from Butler co., Iowa,) reached Denver, March 25th; drove the first wagon to these diggings; have been here five weeks; worked first on a claim, on which they ran a sluice but one day;

produced $225; sold their claim for $2500; are now working a claim on the Hunter lead, have only sluiced one, (this) day; three men employed; produced $85.

Bates & Co., one sluice, run a half day; produced $125.

Colman, King & Co., one sluice, run half a day; produced $75.

Shorts & Collier, bought our claims seven days since of Casto, Kendall & Co., for $2500; $500 down, and the balance as fast as taken out. Have not yet got our sluices in operation. Mr. Dean, from Iowa, on the 6th inst., washed from a single pan of dirt taken from the claim, $17.80. Have been offered $10,000 for the claim.

S. G. Jones & Co., from eastern Kansas, have run our sluices two days, with three men; yield $225 per day. Think the quartz generally in this vicinity is gold-bearing. Have never seen a piece crushed that did not yield gold.

A. P. Wright & Co., from Elkhart co., Ind. Sluice but just in operation; have not yet ascertained its products. Our claim prospects from 25 cents to $1.25 to the pan.

John H. Gregory, from Gordon co., Georgia. Left home last season, en route for Frazier river, was detained by a succession of accidents at Ft. Laramie, and wintered there. Meanwhile, heard of the discoveries of gold on the South Platte, and started on a prospecting tour on the eastern slope of the Rocky mountains in early January. Prospected in almost every valley, from the Cache la Poudre creek, to Pike's Peak, tracing many streams to their sources.– Early in May arrived on Clear creek, at the foot of the mountains, 30 miles southeast of this place. There fell in with the Defrees & Zeigler Indiana companies, and William Fouts, of Missouri. We all started up Clear creek, prospecting.

Arrived in this vicinity, May 6; the ice and snow prevented us from prospecting far below the surface, but the first pan of surface dirt, on the original Gregory claim yielded $4.– Encouraged by this success, we all staked out claims, found the "lead" consisting of burnt quartz, resembling the Georgia mines, in which I had previously worked. Snow and ice prevented the regular working of the lead till May 16th.– From then until the 23rd, I worked it five days with two hands, result, $972. Soon after, I sold my two claims for $21,000 the parties buying, to pay me, after deducting their expenses, all they take from the claims to the amount of $500 per week, until the whole is paid. Since that time, I have been prospecting for other parties, at about $200 per day. Have struck another lead on the opposite side of the valley, from which I washed $14, out of a single pan.

Some forty or fifty sluices commenced, are not yet in operation; but the owners inform us that their "prospecting" shows from 10 cents to $5 to the pan. As the "leads" are all found in the hills, many of the miners are constructing trenches to carry water to them, instead of building their sluices in the ravines, and carrying the dirt thither in wagons, or sacks. Many persons who have come here without provisions or money, are compelled to work, at from $1 to $3 per day and board, until they can procure means of sustenance for the time necessary to prospecting; building sluices, etc. Others, not finding gold the third day, or disliking the work necessary to obtaining it, leave the mines in disgust, after a very short trial, declaring there is no gold here in paying quantities. It should be remembered that the discoveries made thus far, are the result of but five weeks' labor.

In nearly every instance, the gold is estimated by the

miners as worth $20.00 per ounce, which, for gold collected by quicksilver, is certainly a high valuation, though this is undoubtedly of very great purity. The reader can reduce the estimate if he sees fit. We have no data on which to act in the premises.

The wall rock is generally shattered, so that it, like the veinstone, is readily taken out with the pick and shovel. In a single instance only did we hear of wallrock too hard for this.

On the veinstone, probably not more than one-half is so decomposed that the gold can be washed from it. The residue of the quartz is shoveled out of the sluices, and reserved to be crushed and washed hereafter. The miners estimate this as equally rich with that which has "rotted" so that the gold may be washed from it; hence, that they realize, as yet, but half the gold dug by them. This seems probable, but its truth remains to be tested.

It should be borne in mind that, while the miners here now labor under many obvious disadvantages, which must disappear with the growth of their experience and the improvement of their now rude machinery, they at the same time enjoy advantages which cannot be retained indefinitely, nor rendered universal. They are all working very near a mountain stream, which affords them an excellent supply of water for washing at a very cheap rate; and, though such streams are very common here, the leads stretch over rugged hills and considerable mountains, down which the veinstone must be carried to water, at a serious cost. It does not seem probable that the thousands of claims already made or being made on these leads can be worked so profitably in the average as those already in operation. We hear already of many who have worked their claims for days (by panning) without having "raised the color," as the phrase is – that is, without having found any gold whatever. We presume thousands are destined to encounter lasting and utter disappointment, quartz veins which bear no gold being a prominent feature of the geology of all this region.

We cannot conclude this statement without protesting most earnestly against a renewal of the infatuation which impelled thousands to rush to this region a month or two since, only to turn back before reaching it, or to hurry away immediately after more hastily than they came. Gold-mining is a business which eminently requires of its votaries capital, experience, energy, endurance and in which the highest qualities do not always command success. There may be hundreds of ravines in these mountains as rich in gold as that in which we write, and there probably are many; but, up to this hour, we do not know that any such have been discovered. There are said to be five thousand people already in this ravine, and hundreds more pouring into it daily. Tens of thousands more have been passed by us on our rapid journey to this place, or heard of as on their way hither by other routes. For all these, nearly every pound of provisions and supplies of every kind must be hauled by teams from the Missouri river, some 700 miles distant, over roads which are mere trails, crossing countless unbridged water courses, always steep-banked and often mirey, and at times so swollen by rains as to be utterly impassable by waggons. Part of this distance is a desert, yielding grass, wood, and water only at intervals of several miles, and then very scantily. To attempt to cross this desert on foot is madness – suicide – murder. To cross it with teams in mid-summer when the water-courses are mainly dry, and the grass eaten up, is possible only to those who know just where to look for grass and water, and where water must be carried along to preserve life. A few months hence –

probably by the middle of October,– this whole Alpine region will be snowed under and frozen up, so as to put a stop to the working of sluices if not to mining altogether. There then, for a period of six months, will be neither employment, food, nor shelter within five hundred miles for the thousands pressing hither under the delusion that gold may be picked up here like pebbles on the sea-shore, and that when they arrive here, even though without provisions or money, their fortunes are made. Great disappointment, great suffering, are inevitable; few can escape the latter who arrive at Denver City after September without ample means to support them in a very dear country, at least through a long winter. We charge those who manage the telegraph not to diffuse a part of our statement without giving substantially the whole; and we beg the press generally to unite with us in warning the whole people against another rush to these gold-mines, as ill-advised as that of last spring – a rush sure to be followed like that by a stampede, but one far more destructive of property and life. Respectfully,

HORACE GREELEY
A. D. RICHARDSON
HENRY VILLARD

[WILLIAM GREEN RUSSELL][472]

Gregory gulch, Kansas, June 17th, 1859

Messrs. Editors: – As it is likely that any news relating to the mining operations in this country, will not be uninteresting to you or your readers, I will en-

472 The *Dahlonega Mountain Signal,* reprinted in the *Georgia Citizen,* August 19, 1859. William Green Russell, of Georgia, was the leader of the party that made the first discoveries on the South Platte in July 1858. The story of this activity is given in the preceding volume (IX) of this *Southwest Historical Series.*

deavor to give you a few statistics relative to what is
going on in our immediate neighborhood:

The prospects in the veins or mountain diggings, as
they call them here, is improving. There are now about
75 sluices in operation, all washing the ore taken from
the veins, which is paying variously; from $5 to $50 per
day to the hand. Some few are prepared and haul their
dirt, but most of them put it in sacks and carry it to the
branch or creek. New discoveries are being made every
day. There is a company of five men working within
two hundred yards of this place, on a vein which they
lately discovered, and yesterday they carried the ore to
the branch, washed it in a common box-sluice, and
made $125; and to-day they have taken $5 out of a
single pan of ore. The gulches and ravines are as yet
but very little worked or prospected. We commenced
operations here last Saturday with one tom or sluice,
five hands and here is the result of the five day's labor:

Saturday,	June	12	.	.	.	.	260 dwts.
Monday	"	14	.	.	.	.	360 "
Tuesday	"	15	.	.	.	.	274 "
Wednesday	"	16	.	.	.	.	140 "
Thursday	"	17	.	.	.	.	202 "

There are several other companies working on the
same gulch, and are generally making fair wages. John
B. Graham and company, of Dawson, and McClusky,
of Hall county, are working on the same gulch, making
at present about an ounce per day to the hand. There
are several other gulches in this vicinity that are expected to pay when worked equally as well as this one.

Quite a number of accidents have lately happened –
several persons have been shot through awkwardness or
carelessness. William Herbert, of Georgia, shot him-

self when at the base of the mountain near Clear creek, and died instantly. Quite a number have been drowned trying to go down the Platte in badly constructed boats or bateaus. Five persons were frozen to death on the divide between the Platte and Arkansas, during a snow storm on the 24th of June. A number of persons who started to cross the plains early this spring, on foot and without money (or brains), have died of starvation.

It is reported here that the bodies of eighteen men have been found in the mountains, who were burned to death by the burning of the pine forest of the mountains. The forests are still burning and will in all probability continue to burn for some time.

I am going to start out to-morrow on a prospecting tour of eight or ten days, and if anything should be developed during that time worthy of notice I will let you know it. Very respectfully, WILLIAM G. RUSSELL.

[GEORGE M. WILLING, JR.] [473]

At the base of the Snowy range of the Rocky mountains, west of the Gregory gold diggings, 40 miles east of the Middle Park, and 150 miles of the head of Grand river, a tributary of the Colorado of the West.

Tuesday, June 21, 1859.

. . . I have only [recently] visited the Gregory mining district of the mountains, on the north fork of Clear creek, a description of the character of which I will here attempt to give, together with the richness of the lodes discovered and future prospects. At present there are about five hundred sluices in operation, with varied success in the precious ore. The sluices on the Gregory lead, for which the company paid $21,000, are taking

---

473 *Missouri Republican*, July 11, 1859. For additional data on Willing's activities see "Letters of George M. Willing, 'Delegate of Jefferson Territory,'" edited by L. R. Hafen, in *Colorado Magazine*, XVII, 184-189.

out daily $1,000 and upwards. Other sluices are pay-
ing from ten cents to two dollars and half per pan; the
Burroughs lode pays sixty-nine cents on an average to
the pan, in making a hundred pans. The time these
sluices have been in operation have been as follows:
Four sluices on the Gregory lode, three weeks; the Russell Georgia company, same time; the Autrichon (of
Florisant, St. Louis), ten days, pays $80 a day; the
South Bend, Indiana, company's sluices, three weeks,
pays from $300 to $450 a day; other claims that have
not been fairly opened, and who are washing the dirt
from the top of the mountains, are making from
$2[.]oo to $20[.]oo a day to the hand; the Illinois company are making large wages. I have examined the
mammoth quartz lode, and find it to be very rich; this is
partially decomposed. The Gregory lode is the richest
yet discovered. Now, there is no exaggeration in this
statement, and here I will remark that although I look
upon this small area of the mountains as the richest
gold field I ever saw or read of, I do not wish it to be
understood by the masses, that there are leads for all
who come here equally as rich for them, but I will tell
them in plain terms, that the gold is here, but so erratic
is its disposition in the mountains that it takes energy,
perseverance, privation, and capital to work it out. . .
GEORGE M. WILLING, JR.

_____

[THOMAS L. GOLDEN]⁴⁷⁴

I am in the mines called Gregory's diggings, 30 miles
in the mountains on a small tributary stream of Vasquers fork, about 42 miles west of where Cherry creek
empties into Platte river. We are working in what is
called "leads" running through the mountains, these

_____

⁴⁷⁴ From a letter of June 23, addressed to G. G. Gillette, postmaster of
Nebraska City, in *Nebraska City News*, July 16, 1859.

"leads" are among the quartz rock and average two feet in width and are from one to three miles in length. Some men are here taking out three hundred dollars to the sluice, others not so much; it is reported among the miners here that the Illinois company is taking out to the sluice an average of five hundred dollars. The men here are generally satisfied to stay and work. There are a great many purchasing claims on these heavy "leads" and pay weekly as they take it out. They generally make a contract to pay half they take out every week until the claim is paid for. The thoughts of climbing through the Rocky mountains 30 miles to get to the mines sends a great many back after they reach the base of the mountains. All we have to say to the returning emigrant to the states is to stay in the states, and we will bring the gold there. We ask no one to come here and would have been glad had they stayed at home. We were getting our supplies from [New] Mexico before the spring emigration got here and were satisfied that we would make our fortune by fall, and return to the states, but the men that has been humbuged so are crowding us now and in fact are making the most money. Yours, THOS. L. GOLDEN.

THE SOUTHWEST HISTORICAL SERIES

Comprises the following volumes printed direct from the original unpublished manuscripts, and rare originals

Volume I. (1844-1847)
Webb (James Josiah), Adventures in the Santa Fé Trade, 1844-1847
From the original unpublished manuscripts

Volume II. (1854-1861)
Bandel (Eugene). Frontier Life in the Army, 1854-1861; translated by Olga Bandel and Richard Jente.
From the original unpublished manuscripts

Volume III. (1846-1847)
Gibson (George Rutledge). Journal of a Soldier under Kearny and Doniphan, 1846-1847
From the original unpublished manuscript

Volume IV. (1846-1848)
Marching with the Army of the West, 1846-1848. Comprising the following original unpublished journals:
Johnston (Abraham Robinson) Journal of 1846
Edwards (Marcellus Ball) Journal of 1846-1847
Ferguson (Philip Gooch) Diaries, 1847-1848
Muster roll of Company D, First Regiment of Missouri Mounted Volunteers, June, 1846

Volume V. (1849)
Southern Trails to California in 1849
Early news of the gold discovery
Advertising southern trails and routes
Routes through Mexico to California
Routes from Texas to the gold mines
Routes through Arkansas and along the Canadian
The Cherokee Trail
The Santa Fé Trail
From various contemporary newspapers

Volume VI. (1846-1850)
Garrard (Lewis H.). Wah-To-Yah and the Taos Trail.
From the rare original edition of 1850, with extensive additions by the editor

Volume VII. (1846-1854)
Exploring Southwestern Trails, 1846-1854
Cooke (Philip St. George) Journal of the march of the Mormon Battalion, 1846-1847
From the original unpublished manuscript
Whiting (William Henry Chase) Journal of 1849
From the original unpublished manuscript
Aubry (François Xavier) Diaries of 1853 and 1854
From the original published 1853, in the *Santa Fe Weekly Gazette*

Volume VIII. (1687-1873)
McCoy (Joseph G.) Historic Sketches of the Cattle Trade of the West and Southwest
From the rare original edition of 1874, with extensive additions by the editor

Volume IX. (1859)
Pike's Peak Gold Rush Guidebooks of 1859
Tierney (Luke) History of the Gold Discoveries on the South Platte river; and a guide of the route by Smith & Oaks
Parsons (William B.) The new Gold Mines of Western Kansas: a complete description of the newly discovered gold mines, different routes, camping places, tools and

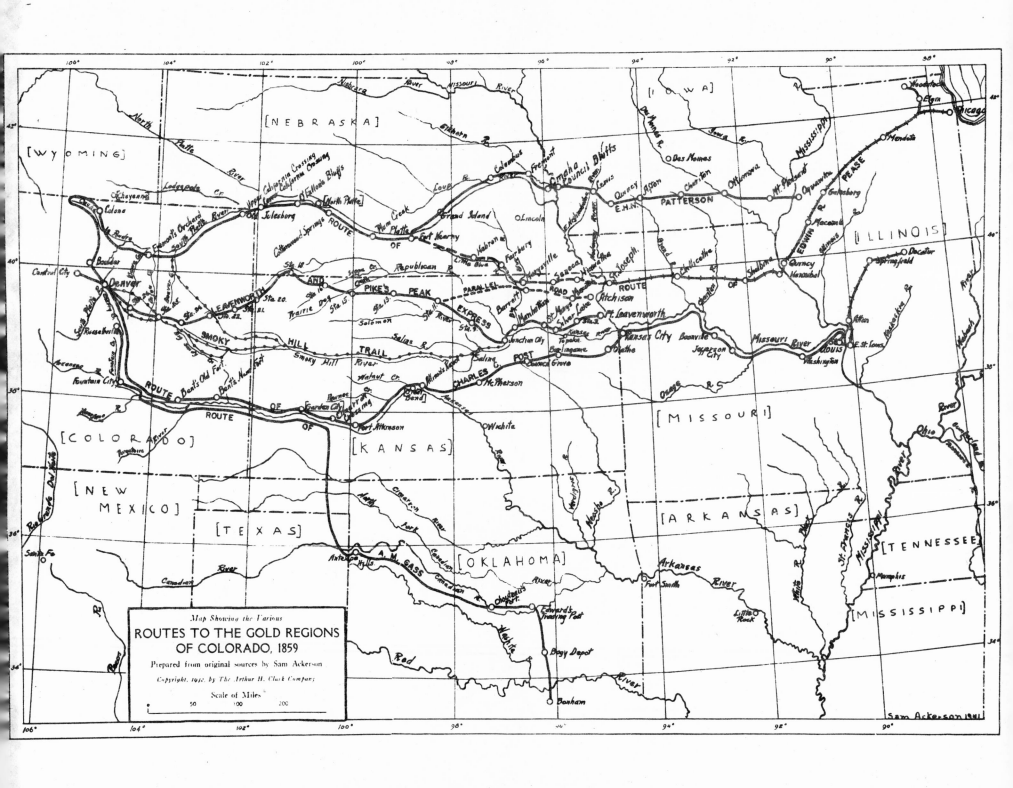

Map Showing the Various
ROUTES TO THE GOLD REGIONS
OF COLORADO, 1859
Prepared from original sources by Sam Ackerson
Copyright, 1941, by The Arthur H. Clark Company
Scale of Miles
50   100   200

Sam Ackerson 1941